STUDY GUIDE
for Yates, Moore, and McCabe's

The Practice of Statistics

TI-83 Graphing Calculator Enhanced

William Notz
Michael Fligner

W. H. Freeman and Company
New York

ISBN 0-7167-4237-3

Printed in the United States of America

First printing 2000

CONTENTS

PREFACE

We have written this study guide to help you learn and review the material in the textbook. The structure of this study guide is as follows. We first provide an overview of each section, which reviews the key concepts of that section. After the overview, there are *guided* solutions to selected problems in that section, along with the key concepts from the section required for solving each problem. The guided solutions provide hints for setting up and thinking about the exercise, which should help improve your problem-solving skills. When you have worked through the guided solution, you can look up the complete solution provided later in the study guide to check your work.

What is the best way to use this study guide? Part of learning involves doing homework problems. In doing a problem, it is best for you to first try and solve the problem on your own. If you are having difficulties, try to solve the problem using the hints and ideas in the guided solution. If you are still having problems, you can read through the complete solution. Generally, try to use the complete solution as a way to check your work. Be careful not to confuse reading the complete solutions with doing the problems themselves. This is the same mistake as reading a book about swimming and believing you are prepared to jump into the deepest end of the pool. If you simply read our complete solutions and convince yourself that you could have worked out the problem on your own, you might be misleading yourself and could have trouble on exams. When you are having difficulty with a particular type of problem, find similar problems to work out in the exercises in the text. Problems adjacent to each other in the text often use the same ideas.

In the overviews, we try to summarize the material that is most important, which should help you review material when you are preparing for a test. If any of the terms in the overview are unfamiliar, you probably need to go back to the textbook and reread the appropriate section. If you are having difficulties with the material, see your instructor for help. Face-to-face communication with your instructor is the best way to clear up difficulties.

CHAPTER 1

EXPLORING DATA

SECTION 1.1

OVERVIEW

Understanding data is one of the basic goals in statistics. To begin, identify the **individuals** or objects described, then the **variables** or characteristics being measured. Once the variables are identified, you need to determine whether they are **categorical** (the variable just puts individuals into one of several groups) or **quantitative** (the variable takes meaningful numerical values for which arithmetic operations make sense).

After looking over the data and digesting the story behind it, the next step is to examine the main features of the data. This process is often referred to as **exploratory data analysis**. The first methods are simple graphs that give an overall sense of the pattern of the data. Choosing the appropriate graphs depends on whether or not the data are numerical. Categorical data (nonnumerical data) use **bar charts** or **pie charts** to summarize the **distribution** of a variable. For quantitative data (numerical data), the distribution can be summarized by using **dotplots**, **histograms**, or **stemplots**. Also, when numerical data are collected over time, in addition to a histogram or stemplot, a **timeplot** can be used to look for interesting features of the data.

When examining the data through graphs, be on the alert for

- **outliers** (unusual values) that do not follow the pattern of the rest of the data;

- some sense of a **center** or typical value of the data;
- some sense of how **spread** out or variable the data are; and
- some sense of the **shape** of the **overall pattern**.

In timeplots, be on the lookout for **trends** over time. Remember that although many of the graphs and plots may be drawn by a computer, it is still up to you to recognize and interpret the important features of the plots and the information they contain.

GUIDED SOLUTIONS

Exercise 1.5

KEY CONCEPTS: Describing histograms – shape and center

a) Look at the overall pattern without focusing on minor irregularities (small ups and downs in the heights of the bars). Find the approximate center and see whether the right and left sides seem similar (symmetric), or whether one side tends to extend farther out than the other (skewed to the right or left).

b) The percent total return of a "typical" stock is described by the center of the histogram. This is the value with about half the observations taking smaller values and the other half taking larger values. You can form a rather accurate guess from visual inspection of the histogram or you can use the following more formal approach. Because the histogram uses percents as the vertical axis, you need to find the percent return with about 50% of the returns below it. This can be done by adding up the heights of the histogram bars starting at the left. It will not be possible to get a sum exactly equal to 50%, but this method will allow you to determine which interval contains the center. This is the best we can do with the information provided in a histogram. (Note that in a stemplot the values of all the observations are given so we could find the central observation exactly.)

Interval containing the center___________Approximate center____________

c) Unlike a stemplot, we cannot determine the minimum and maximum exactly from a histogram. To approximate the highest and lowest returns, first find the bar containing the highest and lowest returns. We know the lowest return is between –70% and –60%. With no further information, we could give –70% as the lowest return or simply indicate that the lowest return is between –70% and –60%.

Minimum ____________________ Maximum ____________________

d) You will need to add the percents for several histogram bars to get to the answer.

Exercise 1.8

KEY CONCEPTS: Drawing a histogram, comparing two histograms

Hints to remember in drawing a histogram:

1. Divide the range of values of the data into classes or intervals of equal length. (Notice that the data in the table are already presented this way.)

2. Count the number of values of our data that fall into each interval. (Again, the table has provided this step.)

3. Draw the histogram.

    a. Mark the intervals on the horizontal axis and label the axis. Include the units.

    b. Mark the scale for the counts or percents on the vertical axis. Label the axis.

    c. Draw bars, centered over each interval, up to the height equal to the count or percent. There should be no space between the bars (unless the count for a class is zero, so that its bar has height zero).

For the exercise:

a) The vertical axis in a histogram can be either the count or percent of the data in each interval. Changing the units from counts to percents will not affect the shape of the histogram. However, to compare two histograms based on different numbers of observations, using percents makes it simpler to compare the histograms by making the scales comparable. Complete the table below of the percent of the total population in each age group. The entry 19.4 given in the table was obtained by dividing 29.3 million, the number under 10 years in 1950, by 151.1 million, the total population in 1950, and multiplying by 100 to give a percent.

AGE DISTRIBUTION (PERCENTAGES), 1950 AND 2075

Age group	1950	2075
Under 10 years	19.4	
10 to 19 years		
20 to 29 years		
30 to 39 years		
40 to 49 years		
50 to 59 years		
60 to 69 years		
70 to 79 years		
80 to 89 years		
90 to 99 years		
100 to 109 years		
Total		

b), c) Complete the two histograms given. Because the values of the variable year have gaps, to avoid gaps in the histograms, the bases of the bars need to be extended to meet halfway between two adjacent values. For example, the bar representing 30 to 39 years would need to meet the bar from 20 to 29 years at 29.5 Similarly, the bar representing 30 to 39 years would need to meet the bar from 40 to 49 years at 39.5. Thus the bar representing 30 to 39 years must go from 29.5 to 39.5, with a center at 34.5. In the histograms given, the *centers* of the bars are provided on the horizontal axis, and the first bar for the 1950 age distribution has been completed for you.

Histogram of age (1950)

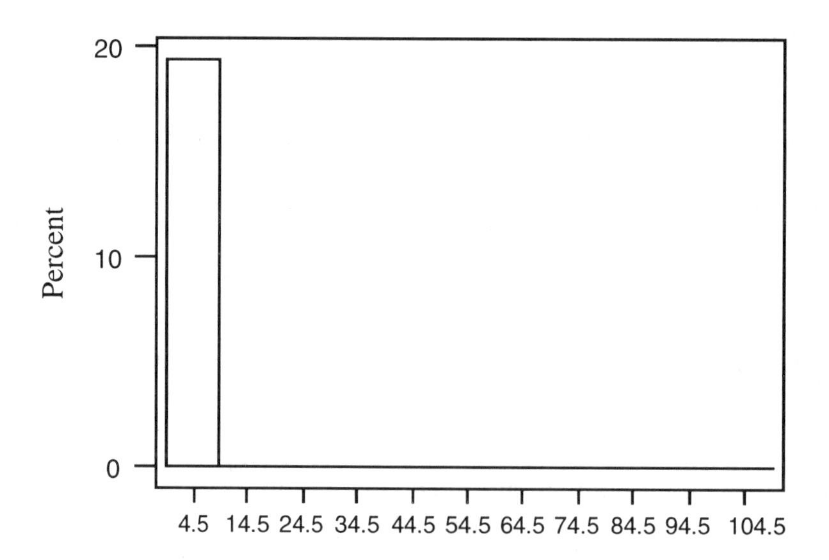

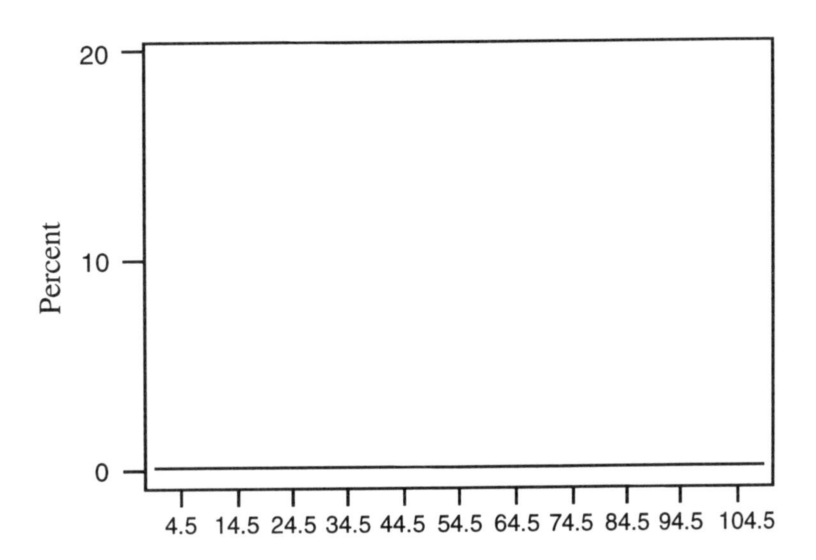

Describe the shapes of the two histograms. Are they approximately symmetric or are they skewed? What are their centers, and what can you say about the spread of each? By comparing the two histograms, what are the most important changes in age distribution between 1950 and 2075?

Exercise 1.13

KEY CONCEPTS: Drawing and interpreting a timeplot

a) Complete the timeplot on the graph below. To get you started, the total fatalities in 1986 and 1987 are included in the plot. Make sure to label the axes and include a title with the timeplot. What is the general pattern in the plot? Does there appear to be a trend? If you have found a trend, can you give some possible reasons for it?

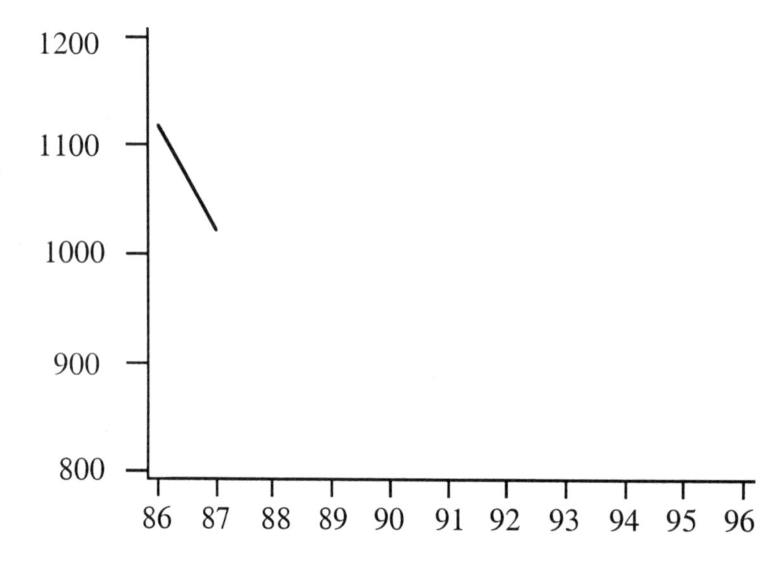

b) Complete the timeplot on the graph below. What is the general pattern in the timeplot? Does there appear to be a trend? If you have found a trend, can you give some possible reasons for it?

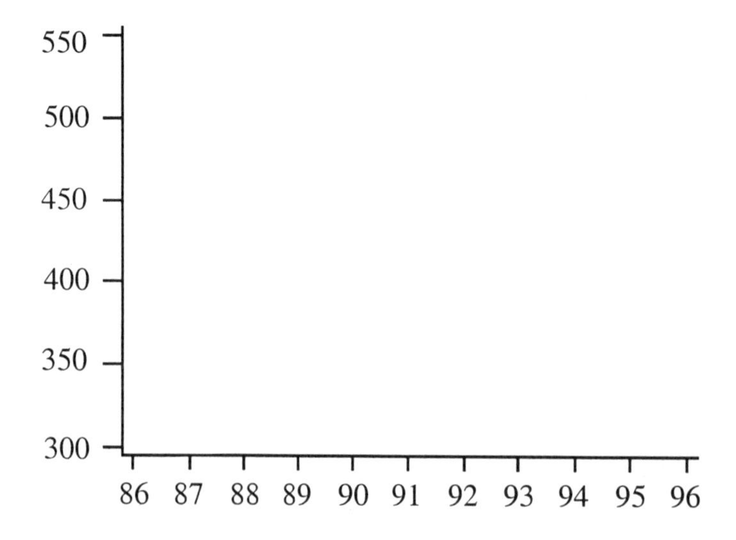

Exercise 1.17

KEY CONCEPTS: Drawing stemplots, back-to-back stemplots, comparing two stemplots

Hints to remember for drawing a stemplot:

1. Put the observations in numerical order.

2. Decide how the stems will be shown. Commonly, a stem is all digits except the rightmost. The leaf is then the rightmost digit.

3. Write the stems in increasing order vertically. Write each stem only once. Draw a vertical line next to the stems.

4. Write each leaf next to its stem.

5. Rewrite the stems and put the leaves in increasing order.

For the exercise:
a) We have given the stems below. The stems are in units of ten home runs. We have included one observation to get you started.

```
2 | 2
3 |
4 |
5 |
6 |
```

Is the distribution symmetric, skewed, or neither? About how many home runs did Ruth hit in a typical year? (Is this measured by the center or the spread?) Is his record 60 home runs an outlier?

b) We have given the stems below, enclosed in vertical lines. The stems are again in units of ten home runs. Make the stemplot for Ruth in the usual way to the right of the stems. The stemplot for Maris is made in the same way except the leaves go off to the left instead of the right. This type of plot is excellent for comparing two small data sets. Complete the back-to-back stemplot below. One value for Maris and one value for Ruth have been included to help you get started.

Maris		Ruth
8	0	
	1	
	2	2
	3	
	4	
	5	
	6	

How do the two players compare? Concentrate on center, spread, outliers, and overall shape of the two stemplots. How does your plot show Ruth's superiority as a home run hitter?

COMPLETE SOLUTIONS

Exercise 1.5

a) Taking the center at about 10–20% (see part b), the shape of the histogram is fairly symmetric. There is no apparent skewness in either direction. (You might view this as slightly skewed to the right, but the skewness is quite minimal). Approximately half the stocks had returns between 0% and 30% (the sum of the heights of the three central bars).

b) Visually, any guess between about 10% and 20% is reasonable. The first 8 bars of the histogram add to a total of 1 + 1 + 1 + 1 + 3 + 5 + 11 + 16 = 39%, giving 39% of the stocks with returns below 10% (10% is the right endpoint of the eighth interval). Including the next bar gives an additional 21%, bringing the total to 39 + 21 = 60%. Thus the center must occur in the interval corresponding to the ninth histogram bar, which is the interval from 10% to 20%. Lacking additional information, we could use the midpoint of this interval, which gives a return of 15% as the approximate center.

c) The smallest total return is about –70% (the left endpoint corresponding to the bar containing the smallest value) and the largest total return is about 110% (the right endpoint corresponding to the bar containing the largest value). Thus the spread of the distribution is from about –70% to 110%.

d) You need to add the heights of the first seven bars in the histogram, because the intervals corresponding to each of these bars correspond to negative total returns. Thus 1 + 1 + 1 + 1 + 3 + 5 + 11 = 23% of the stocks lost money.

Exercise 1.8

a)

Age group	1950	2075
Under 10 years	19.4	11.2
10 to 19 years	14.4	11.5
20 to 29 years	15.9	11.8
30 to 39 years	15.1	12.3
40 to 49 years	12.8	12.2
50 to 59 years	10.3	12.1
60 to 69 years	7.3	11.1
70 to 79 years	3.6	8.8
80 to 89 years	1.1	6.1
90 to 99 years	0.1	2.5
100 to 109 years	0.0	0.5
Total	100.0	100.1

b), c) The histograms are reproduced below. They use the same numerical scale as in the guided solutions, but the numbers have been suppressed.

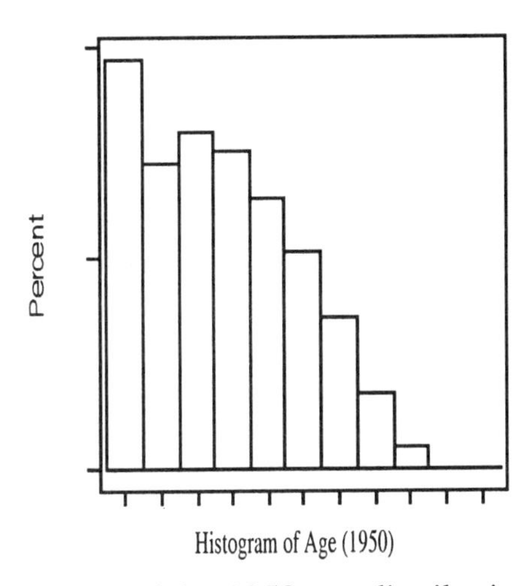

Histogram of Age (1950)

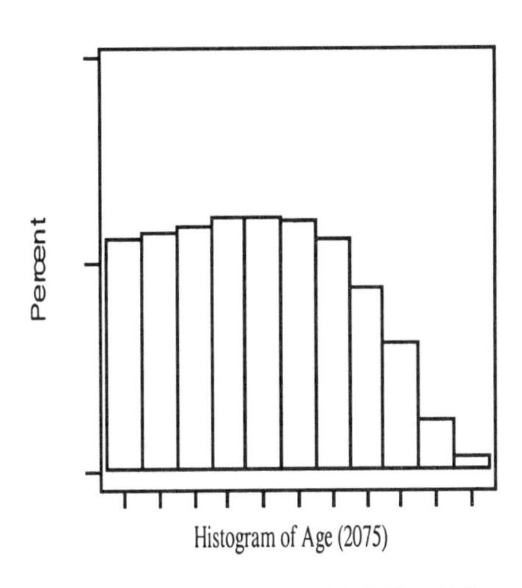

Histogram of Age (2075)

The center of the 1950 age distribution is pretty close to 30 years—49.7% of the people are up to 29 years old (the first three age intervals) with the remaining 50.3% above 29. In this case, the calculation is most easily done by adding the percentages given in part a rather than trying to read them off the histogram. For the projected 2075 age distribution, the center is in the 40– to 49–year age interval, probably close to the midpoint or about 44 years. We can't specify it more exactly from the information given. The 1950 age distribution is skewed to the right, and has much less spread than the projected 2075 age distribution. The shape for the projected 2075 age distribution has the ages pretty evenly spread up to 70 or 80 years (the histogram is pretty flat until this point because the percents are similar for these intervals), with a small percentage (just under 10%) exceeding 80 years.

Exercise 1.13

Total number of Virginia road fatalities, 1986–1996

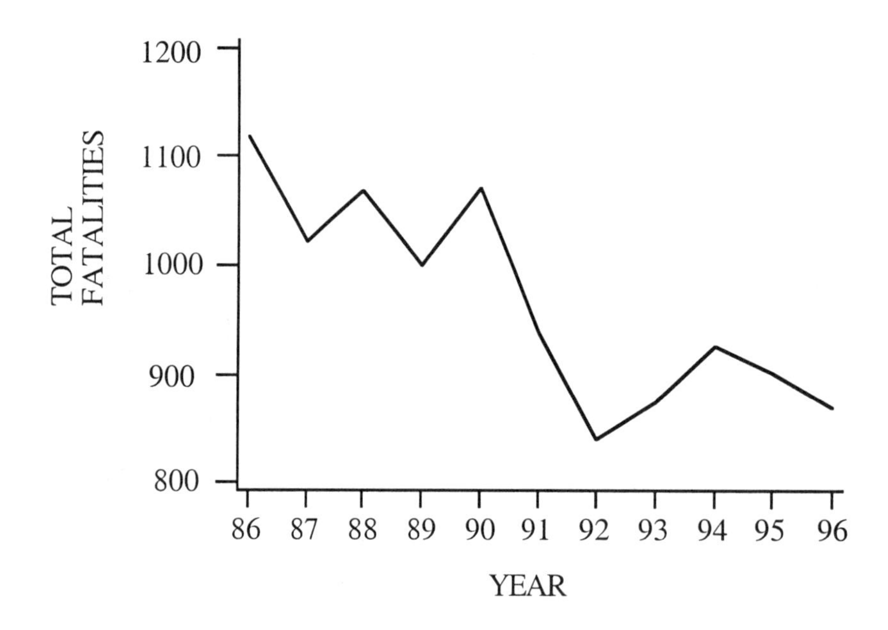

Total number of alcohol related road fatalities, 1986–1996

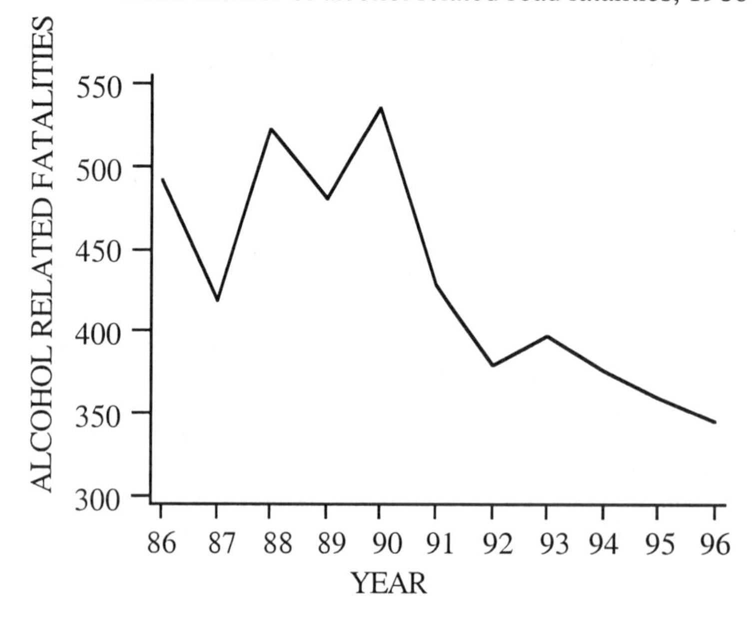

a) There is a general downward trend in the number of fatalities over the ten-year period. (Note: the number of fatalities, not the number of accidents is being plotted). Possible reasons might include improved safety in cars, airbags and mandatory use of seat belts, and more public awareness of driver safety. From the graph, it appears that most of the decrease occurred around 1990, with the lower fatality levels persisting for the next five years of data.

b) There is a general downward trend in the number of alcohol-related road fatalities over the ten-year period. Possible reasons might include those in part a plus greater awareness and intolerance of drunk driving, as well as stiffer penalties for drunk driving.

Exercise 1.17

a)

2	25
3	45
4	1166679
5	449
6	0

Ruth's distribution is centered at 46 home runs per year (the center describes the "typical" number of home runs hit), and looks fairly symmetric. His famous record of 60 home runs in 1927 does not appear to be an outlier given his record.

b)

Maris		Ruth
8	0	
643	1	
863	2	25
93	3	45
	4	1166679
	5	449
1	6	0

The distribution of Maris is centered at 23 home runs per year with the record of 61 home runs for Maris clearly being an outlier. From the back-to-back stemplots, the superiority of Ruth as a home run hitter is clearly evident. Ruth hit more than 40 home runs in 11 of his 15 years, while Maris did this only in his record year. Maris also hit fewer than 20 home runs in 4 of his 10 years, and Ruth never hit this few home runs in a season.

SECTION 1.2

OVERVIEW

Once you examine graphs to get an overall sense of the data, numerical summaries of features of the data make the notions of center and spread more precise.

Measures of center:
- **mean** (often written as $\bar{x}$)
- **median** (often written as M)

Finding the mean, $\bar{x}$.

The mean is the common arithmetic average.

If there are n observations, $x_1, x_2, \ldots, x_n$, then the mean is

$$\bar{x} = \frac{x_1 + x_2 + \ldots + x_n}{n} = \frac{1}{n}\sum x_i$$

Recall that Σ means "add up all these numbers."

Finding the median, M.

1. List all the observations from the smallest to the largest.

2. If the number of observations is odd, then the median is the middle observation. Just count from the bottom of the list of ordered values up to the $(n + 1)/2$ largest observation. This observation is the median.

3. If the number of observations is even, then the median is the average of the two center observations.

Measures of spread:
- **quartiles** (often written as Q_1 and Q_3)
- **interquartile range** (IQR)
- **standard deviation** (s)
- **variance** (s^2)

Finding the quartiles, Q_1 and Q_3.

1. Locate the median.

2. The first quartile, Q_1, is the median of the lower half of the list of ordered observations.

3. The third quartile, Q_3, is the median of the upper half of the list of ordered values.

Finding the interquartile range, *IQR*.

1. First find the quartiles Q_1 and Q_3.

2. The interquartile range is the difference, $Q_3 - Q_1$.

Finding the variance, s^2, and the standard deviation, s.

1. Take the average of the squared deviations of each observation from the mean. In symbols, if we have n observations, $x_1, x_2, \ldots, x_n$, with mean $\bar{x}$,

$$s^2 = \frac{(x_1 - \bar{x})^2 + (x_2 - \bar{x})^2 + \ldots + (x_n - \bar{x})^2}{n-1} = \frac{1}{n-1}\sum (x_i - \bar{x})^2$$

(Remember, Σ means "add up.")

If you are doing this calculation by hand, it is best to take it one step at a time. First calculate the deviations, then square them, sum them up, and divide the result by $n-1$.

2. The standard deviation is the square root of the variance, i.e., $s = \sqrt{s^2}$. Here are some things to remember about the standard deviation:

a. s measures the spread around the mean.

b. s should be used only with the mean, not the median.

c. If $s = 0$, then all the observations must be equal.

d. The larger s is, the more spread out the data are.

e. s can be strongly influenced by outliers, and it is best to use s and the mean only if the distribution is symmetric or nearly symmetric.

For measures of spread, the quartiles are appropriate when the median is used as a measure of center. In fact, the **five-number summary**, reporting the largest and smallest values of the data, the quartiles and the median, provides a nice, compact description of the data. The five-number summary can be represented graphically by a **boxplot**. A **modified boxplot** uses the *IQR* to identify outliers and plots these outliers as isolated points on the boxplot.

If you use the mean as a measure of center, then the standard deviation and variance are the appropriate measures of spread. Watch out because means and variances can be strongly affected by outliers and are harder to interpret for skewed data. The mean and standard deviation are not **resistant measures**. The median and quartiles are more appropriate when outliers are present or when the data are skewed. The median and quartiles are resistant measures.

GUIDED SOLUTIONS

Exercise 1.29

KEY CONCEPTS: Measures of center, mean and median

The rule for finding the mean is to add the values of all the observations and divide that by the number of observations. How many observations (employees) are there? What is the total of their salaries?

Number of observations =
Total of the observations =
Mean of the observations =

How many salaries are less than the mean? Why do so many of the employees make less than the mean?

Remember, to find the median, first order the observations; making sure to include a salary as many times as it occurs in the data set. If the number of observations is odd, the median is the observation in the center. If the number of observations is even, the median is the average of the two center observations.

Exercise 1.39

KEY CONCEPTS: Boxplots, side-by-side boxplots

Hints for drawing a boxplot:

1. The center box starts at Q_1 and the ends at Q_3.

2. The median is marked by a line in the center of the box.

3. For a modified boxplot, the whiskers are extended to the largest (or smallest) data point that is not an outlier. The outliers are then plotted as isolated points. (In a "nonmodified" boxplot, the whiskers would extend out from the box to the smallest and largest observations.)

For this exercise:
First find the five-number summary for the three types of hot dogs. The boxplot is then just a graphic that uses these five numbers.

Complete the five-number summary for the three types of hot dogs in the table below. The simplest way is to first list the calories in increasing order for each of the three groups. Next, add the boxplots for meat and poultry alongside the boxplot for beef. If you are having difficulty, we have included the details of the computations of the five-number summary for beef hot dogs following the plot.

Five-number summary	Beef	Meat	Poultry
Minimum	111		
First Quartile	140		
Median	152.5		
Third Quartile	178.5		
Maximum	190		

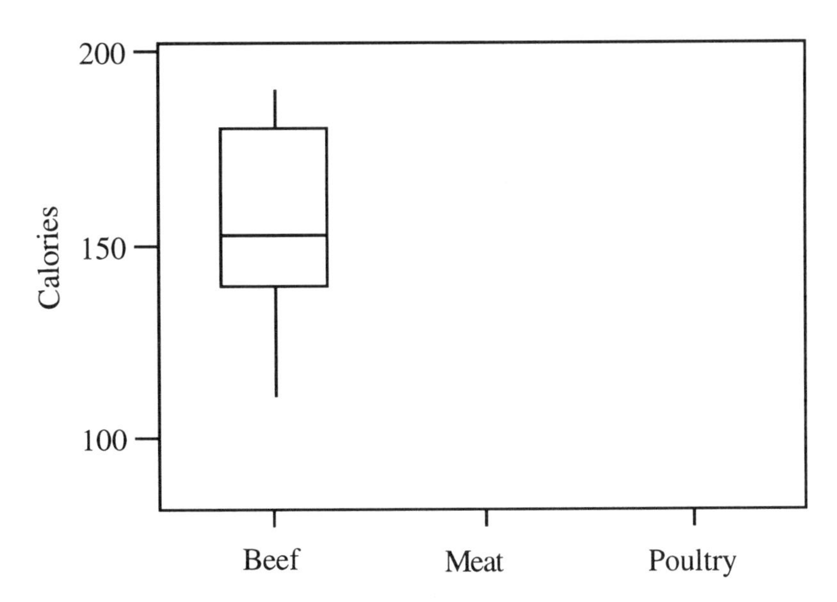

The details of the calculations of the five-number summary for beef hot dogs follow:

First, list the ordered calories for the beef hot dogs:
111 131 132 135 139 141 148 149 149 **152**
153 157 158 175 176 181 184 186 190 190

The minimum and maximum are 111 and 190 calories, respectively. When finding the median and the quartiles, just concentrate on their positions and don't pay attention to the tied values in the data. Because there are 20 brands, the median is the average of the 10th and 11th smallest values, which have been entered in bold face. Since these are 152 and 153, the median is 152.5. There are 10 brands below the median (average of the 10th and 11th) and 10 brands above it. The first quartile is the median of the 10 observations below the median that corresponds to the average of the 5th and 6th smallest observations. These observations are underlined above and their average is 140. The third quartile is the average of the 5th and 6th smallest observations in the second row (those above the median), and their average is 178.5. The interquartile range is $178.5 - 140 = 38.5$ and $1.5 \times IQR = 57.75$. Subtracting $1.5 \times IQR$ from Q_1 and adding it to Q_3, we see that there are no observations labeled as outliers, so the whiskers extend to the smallest and largest observations.

How do the three distributions compare? Will eating poultry hot dogs usually lower your calorie consumption compared to beef or meat?

Exercise 1.41

KEY CONCEPTS: Histograms, stemplots and boxplots, mean and standard deviation

The simplest graph to start with when you have a small data set is a stemplot. The first two stems are given below. Fill in the rest of the stems and complete the stemplot. See Section 1.1 for hints on how to make a stemplot. For these data, a dotplot or a histogram would also be a suitable graph.

```
4.8 | 8
4.9 |
```

Given the distribution of measurements, are $\bar{x}$ and s good measures to describe this distribution? Find their values and give an estimate of the density of the earth based on these measures. Remember, if you are doing the calculation by hand, first find $\bar{x}$ and then calculate s in steps.

Exercise 1.47

KEY CONCEPTS: Standard deviation

There are two points to remember in getting to the answer. First is numbers "further apart" from each other tend to have higher variability than numbers closer together. Second, repeats are allowed. There are several choices for the answer to part a, but only one for part b.

COMPLETE SOLUTIONS

Exercise 1.29

There are five clerks, two junior accountants, and the owner for a total of eight employees. The total of the salaries is

Total salary of clerks	5 × \$22,000	= \$110,000
Total salary of accountants	2 × \$50,000	= \$100,000
Salary of owner		= \$270,000
Total of all salaries		= \$480,000

Number of salaries	8
Mean of the observations	\$480,000/8 = \$60,000

Everyone but the owner makes less than the mean. This is because the mean is not a resistant measure of center and its value is pulled up by the high "outlier" (large salary of the owner). Because prospective employees would usually think of the average salary as a "typical" salary for the firm, a recruiter quoting the average salary as $60,000 without qualifying this statement would mislead most prospective employees into thinking the salaries are higher than they actually are.

The ordered observations are

$22,000 $22,000 $22,000 $22,000 $22,000 $50,000 $50,000 $270,000

The median is the average of the fourth and fifth observations, which is $22,000.

Exercise 1.39

Five-number summary	Beef	Meat	Poultry
Minimum	111	107	86
First Quartile	140	138.5	100.5
Median	152.5	153	129
Third Quartile	178.5	180.5	143.5
Maximum	190	195	170

The ordered calories for the meat hot dogs are as follows:

107 135 136 138 139 140 146 147 **153**
172 173 175 179 182 190 191 195

There are 17 brands and the median is the 9th smallest which has been entered in bold. There are 8 brands below the median and 8 brands above it. The first quartile is the median of the 8 observations below the median, which corresponds to the average of the 4th and 5th smallest observations. These observations are underlined above and their average is 138.5. The third quartile is the average of the 4th and 5th smallest observations in the second row (those above the median) and their average is 180.5. Again, there are no observations labeled as outliers.

The ordered calories for the poultry hot dogs are as follows, and because there are 17 brands, the calculations are identical to the meat hot dogs:

86 87 94 99 102 102 106 113 **129**
132 135 142 143 144 146 152 170

The three side-by-side boxplots comparing the calorie content of the three types of hot dogs are given on the next page.

A comparison of the boxplots for beef and meat shows little difference in these two distributions. However, poultry hot dogs are "generally" lower in calories than either beef or meat. The lower quartile of the poultry distribution is below the minimum number of calories for beef or meat, indicating that about one-fourth of the poultry brands featured in the study had fewer calories than any of the beef or meat brands included in the study.

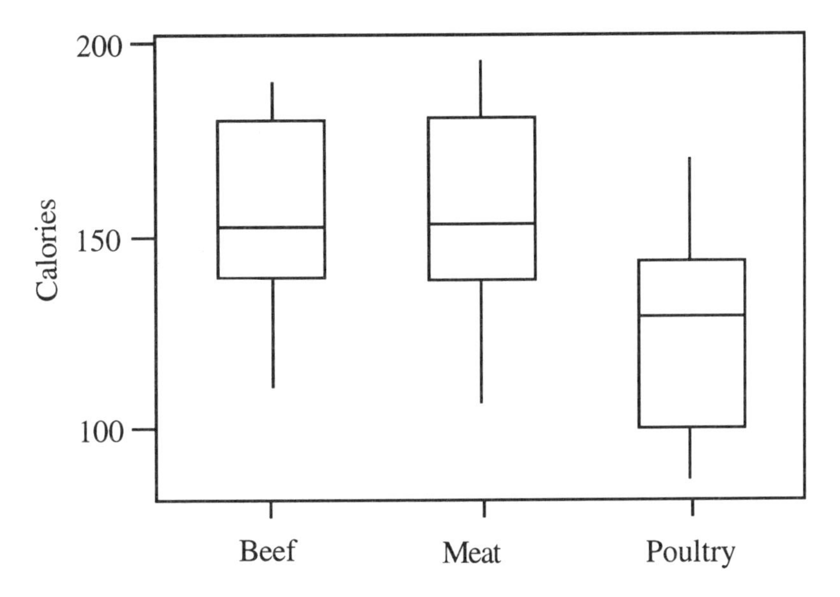

Exercise 1.41

The stemplot gives a distribution that appears fairly symmetic with one low outlier of 4.88, but this is not that far from the bulk of the data. In this case, $\bar{x}$ and s should provide reasonable measures of center and spread.

```
4.8 | 8
4.9 |
5.0 | 7
5.1 | 0
5.2 | 6799
5.3 | 04469
5.4 | 2467
5.5 | 03578
5.6 | 12358
5.7 | 59
5.8 | 5
```

In practice, you will be using software or your calculator to obtain the mean and standard deviation from keyed-in data. However, we illustrate the step-by-step calculations below, to help you understand how the standard deviation works. Be careful not to round off the numbers until the last step, as this can sometimes introduce fairly large errors when computing s.

Observation x_i	Difference $x_i - \bar{x}$	Difference squared $(x_i - \bar{x})^2$
5.50	0.052100	0.002714
5.61	0.162100	0.026277
4.88	-0.567900	0.322510
5.07	-0.377900	0.142808
5.26	-0.187900	0.035306
5.55	0.102100	0.010424
5.36	-0.087900	0.007726
5.29	-0.157900	0.024932
5.58	0.132100	0.017450
5.65	0.202100	0.040845
5.57	0.122100	0.014908
5.53	0.082100	0.006740
5.62	0.172100	0.029618
5.29	-0.157900	0.024932
5.44	-0.007900	0.000062
5.34	-0.107900	0.011642
5.79	0.342100	0.117033
5.10	-0.347900	0.121034
5.27	-0.177900	0.031648
5.39	-0.057900	0.003352
5.42	-0.027900	0.000778
5.47	0.022100	0.000488
5.63	0.182100	0.033161
5.34	-0.107900	0.011642
5.46	0.012100	0.000146
5.30	-0.147900	0.021874
5.75	0.302100	0.091265
5.68	0.232100	0.053870
5.85	0.402100	0.161684
Column Sums = 157.99		1.366869

The mean is 157.99/29 = 5.4479. This has been subtracted from each observation to give the second column of deviations or differences $x_i - \bar{x}$. The second column is squared to give the squared deviations or $(x_i - \bar{x})^2$ in column 3. The variance is the sum of these squared deviations divided by one less than the number of observations:

$$s^2 = \frac{1.366869}{29-1} = 0.0488168 \text{ and } s = \sqrt{0.0488168} = 0.22095.$$

Use the mean of 5.4479 as the estimate of the density of the Earth based on these measurements, because the distribution is approximately symmetric and has no outliers.

Exercise 1.47

a) The standard deviation is always greater than or equal to zero. The only way it can equal zero is if all the numbers in the data set are the same. Because repeats are allowed, just choose all four numbers the same to make the standard deviation equal to zero. Examples are 1, 1, 1, 1 or 2, 2, 2, 2.

b) To make the standard deviation large, numbers at the extremes should be selected. So you want to put the four numbers at zero or ten. The correct answer is 0, 0, 10, 10. You might have thought that 0, 0, 0, 10 or 0, 10, 10, 10 would be just as good, but a computation of the standard deviation of these choices shows that two at either end is the best choice.

c) There are many choices for part a but only one for part b.

SELECTED TEXT REVIEW EXERCISES

GUIDED SOLUTIONS

Exercise 1.54

KEY CONCEPTS: Stemplots, five-number summary, describing a distribution

The stemplot produced by SPLUS and provided in the problem has a slightly different format from others given in the text. If there are outliers, rather than producing the many extra stems required to include the outliers in the stemplot, the computer package lists the high and low outliers separately and does not include them in the actual stemplot. The four low outlying observations are –34.042255, –31.25, –27.06271, and –26.6129. The next three smallest observations are given in the first stem and are –19, –18, and –15. Similarly, at the upper end, there are five high outliers that are separated from the body of the stemplot. Summary statistics also are included in the computer output.

a) All numbers for the five-number summary can be read directly from the output. Remember that the minimum and maximum are included with the outliers in the SPLUS output.

b) Identify the rate of return for the best month. What would $1000 worth of Wal-Mart stock have been worth at the end of this month?

Now do the same for the worst month.

c) What is the numerical value of the *IQR*?

$IQR = Q_3 - Q_1 =$

To identify outliers first compute $1.5 \times IQR =$_______________ and then

$Q_1 - 1.5 \times IQR =$
$Q_3 + 1.5 \times IQR =$

Does it appear that SPLUS uses this criterion to identify outliers?

d) What are the main features of this distribution?

Exercise 1.55

KEY CONCEPTS: Timeplot

Some of the interesting features in these data are the patterns that might occur over time. However, these features are lost when the data are presented in a stemplot. The side-by-side boxplots for the 19 years of data allow you to look for interesting patterns occurring over time.

a) To answer this, look at the medians (centers of the boxes) and see whether you notice any patterns or long-term trends.

b) The spread for a given year can be examined in two ways. The first is to consider the length of the box (from first to third quartile) as a measure of spread, and the second is to use the length from the minimum to the maximum as the measure of spread. The two often give similar impressions. The disadvantage of the second is that it is completely determined by only two observations (the minimum and the maximum), which may give a misleading impression of the spread for the remainder of the data.

c) Examining the minima and the maxima will show you where most of the outliers are. See whether you can determine where some of the remaining ones are by a closer inspection of the boxplots.

COMPLETE SOLUTIONS

Exercise 1.54

a) The minimum is –34.04255 (the smallest low outlier) and the maximum is 58.67769 (the largest high outlier). The first quartile, median, and third quartile are all listed as part of the computer output and are –2.950258, 3.4691, and 8.4511, respectively.

b) The highest rate of return is 58.67769%. A $1000 investment would have earned $1000 x 0.5867769 or $586.68 and would have been worth $1586.68 at the end of the month. The lowest rate of return was –34.04255%. If you had invested in this month, you would have lost $340.43, so your initial investment of $1000 would have been worth $659.57 at the end of this month.

c) $IQR = Q_3 - Q_1 = 8.4511 - (-2.950258) = 11.401358$

To identify outliers first compute 1.5 x IQR = 17.102037 and then

$$Q_1 - 1.5 \times IQR = -2.950258 - 17.102037 = -20.052295$$
$$Q_3 + 1.5 \times IQR = 8.4511 + 17.102037 = 25.553137.$$

It appears that SPLUS uses the $1.5 \times IQR$ criterion to identify outliers since all numbers identified as high and low outliers satisfy this criterion.

d) The distribution is fairly symmetric with a few high and low outliers. The mean and median are fairly close, which is a consequence of the roughly symmetric distribution.

Exercise 1.55

a) The first two years have a median rate of return that is negative. From 1975 onward, the median rate of return was positive (except for 1984), but just fluctuated up and down without showing any long-term trend to either increase or decrease.

b) Looking at either the lengths of the boxes or the distance from the minimum to the maximum, the first three years have much larger spreads than succeeding years. The years from 1976 through 1991 have fairly similar spreads, the only exception being 1987, which has one low outlier.

c) Many of the outliers show up as either the maximum or minimum for a given year. The 58.7 occurred in 1973, and the 57.9 occurred in 1975. Because 1973 to 1975 are the only years with rates of return above 40, the 41.8 and the 42.0 must have occurred in these three years. One of these two values must be the maximum in 1974, although it is not possible to tell which from the graph. The

32 occurred in 1979 (although the value of the maximum in 1979 can't be determined exactly from the graph, the next largest observation is 24 and the maximum in 1979 clearly exceeds this value). The only years with rates of return below –20 were 1973 and 1987. The two lowest outliers, –34 and –31, came from 1973, and at least one of the outliers –27.1 and –26.6 came from 1987, the other possibly from 1973. These observations agree with the conclusion from part b, namely that the greatest spread occurred in the first three years.

CHAPTER 2

THE NORMAL DISTRIBUTIONS

SECTION 2.1

OVERVIEW

This section considers the use of smooth curves to describe the overall pattern of a distribution. A smooth curve that summarizes the shape of a histogram is known as a **density curve**, which is an idealized histogram. The area under a density curve and above any range of values represents the proportion of the data that fall in that range. Like a histogram, it can be described by measures of center such as the **median** (which is a point denoting that half the area under the density curve is to the left of the point) and the **mean** (the center of gravity or balance point of the density curve). To distinguish between the mean $\bar{x}$ and standard deviation s computed from actual observations, we denote the mean of a density curve by μ and the standard deviation of a density curve by σ.

One of the most commonly used density curves in statistics is the **normal curve**. Normal curves are symmetric and bell-shaped. The peak of the curve is located above the mean and median, which are equal because the density curve is symmetric. The standard deviation measures how concentrated the area is around this peak. Normal curves follow the 68–95–99.7 rule, i.e., 68% of the area under a normal curve lies within one standard deviation of the mean (illustrated in the figure below), 95% within two standard deviations of the mean, and 99.7% within three standard deviations of the mean.

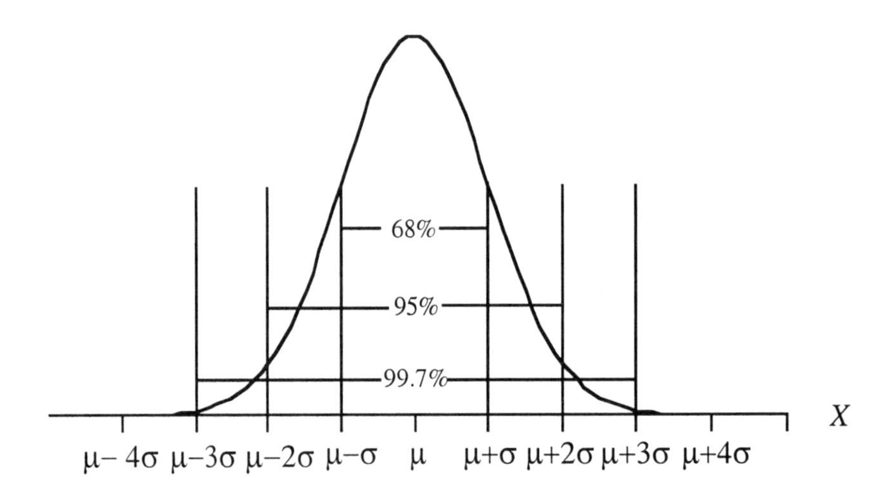

GUIDED SOLUTIONS

Exercise 2.2

KEY CONCEPTS: Density curves and area under a density curve

a) The percent of the observations that lie above 0.8 corresponds to the area under the density curve above 0.8. This area is the shaded region in the figure below. Although not indicated in the figure, the height of the density curve is 1 because the total area must be 1.

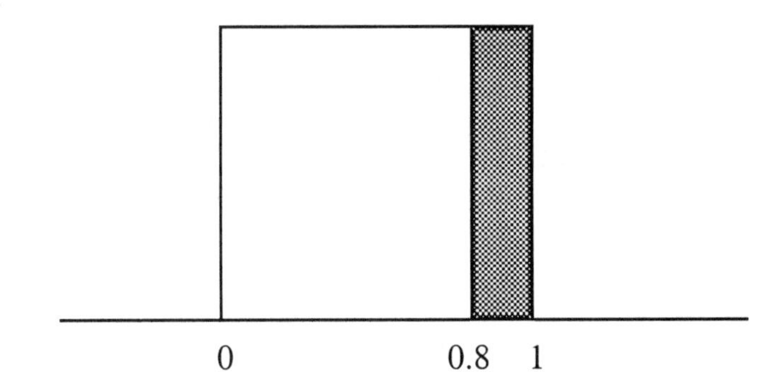

The area between 0.8 and 1 under the density curve forms a rectangle whose base is of length 0.2 and whose height is of length 1. Recall that the area of any rectangle is the product of the length of the base of the rectangle and the height. In this case this product is $0.2 \times 1 = 0.2$.

b) The area of interest is the shaded region in the figure below.

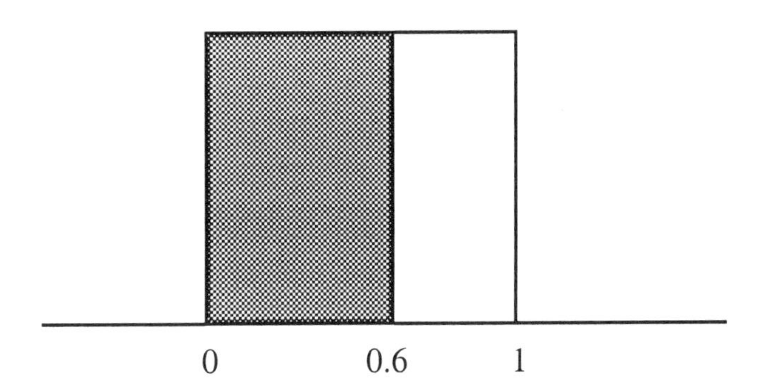

Compute the area of the shaded region by filling in the blanks below.

area = length of base × height = __________ × __________ = __________

c) Try answering this part on your own, using the same reasoning as in part b. First, shade in the area of interest in the figure below.

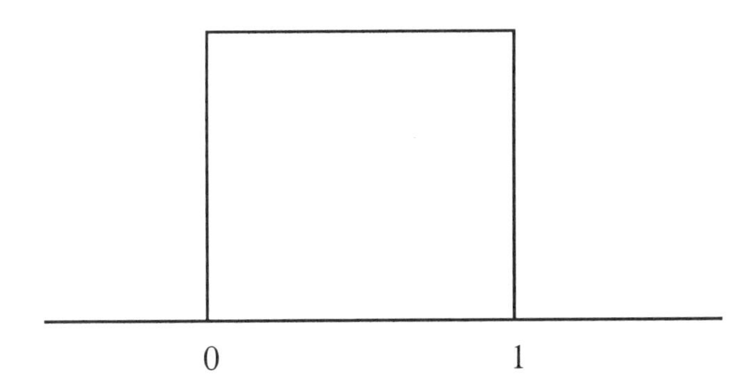

Next, compute the area of your shaded region as in part b.

area = length of base × height = __________ × __________ = __________

Exercise 2.12

KEY CONCEPTS: The 68–95–99.7 rule for normal density curves

Recall that the 68–95–99.7 rule says that the area under a normal curve between the mean minus one standard deviation and the mean plus one standard deviation is 0.68, the area under a normal curve between the mean minus two standard deviations and the mean plus two standard deviations is 0.95, and the area under a normal curve between the mean minus three standard deviations and the mean plus three standard deviations is 0.997. In additon the area under a density curve between two numbers corresponds to the proportion of data that lies between these two numbers.

In this problem, the mean is 22.8 inches and the standard deviation is 1.1 inches. The 68–95–99.7 rule says the following:

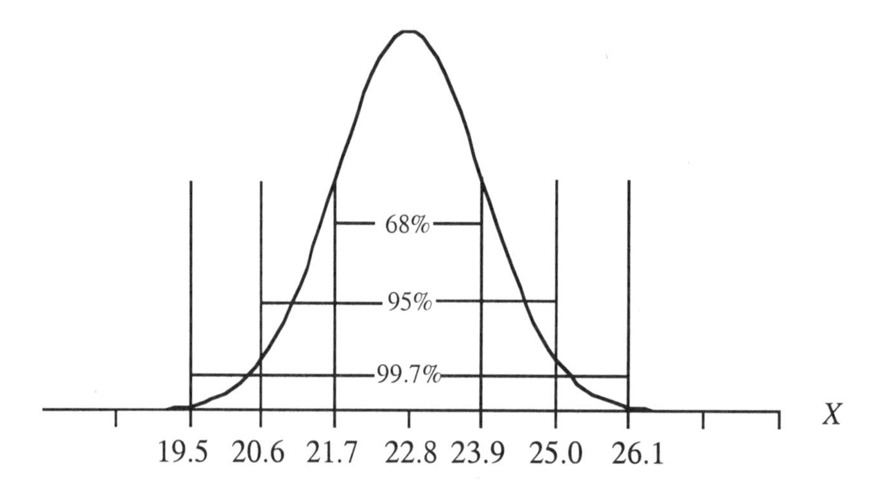

a) From the figure, we see that 68% of soldiers have head circumferences from 21.7 to 23.9 inches. The remaining 32% have head circumferences either smaller than 21.7 inches or larger than 23.9 inches. The symmetry of the normal distribution implies that this 32% will be evenly divided between those with head circumferences smaller than 21.7 inches and those with head circumferences larger than 23.9 inches. Thus, half of 32%, namely 16%, of soldiers have head circumferences greater than 23.9 inches.

b) Recall that the term percentile of a quantity refers to the percentage less than or equal to the quantity. Therefore, we want to find the percent of soldiers having head circumferences less than or equal to 23.9 inches. Based on the answer to part a, what is this percent?

c) Refer to the figure above part a to answer this.

Exercise 2.13

KEY CONCEPTS: The 68–95–99.7 rule for normal density curves

In this problem, the mean is 266 days and the standard deviation is 16 days. According to the 68–95–99.7 rule, the middle 68% of the lengths of all pregnancies, for example, lie between $266 - 1 \times 16 = 250$ and $266 + 1 \times 16 =$ 282 days.

a) What does the 68–95–99.7 rule say about the middle 95% of the lengths of all pregnancies?

b) We indicated above that the middle 68% of all pregnancies have lengths between 250 and 282 days. Thus 100% – 68% = 32% are either below 250 or above 282. What percent must be below 250 (recall that the density curve is symmetric)? What percent must be above 282? The figure below gives the answer.

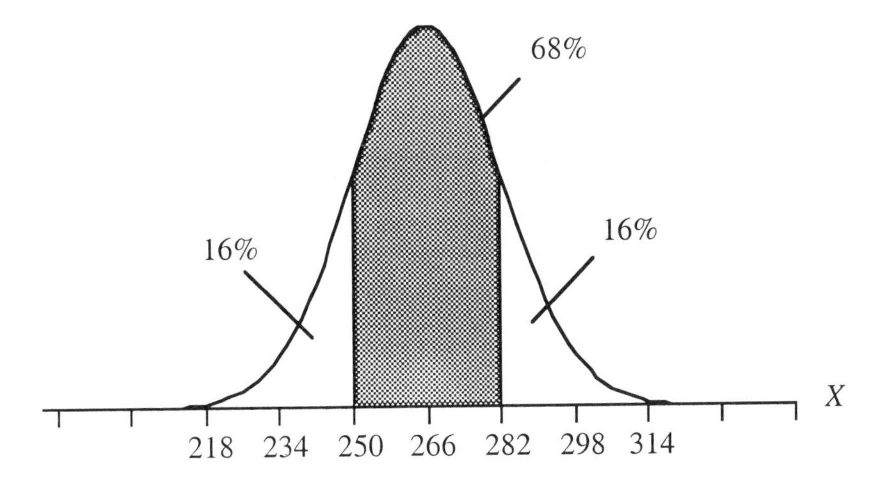

In particular, this tells us that the shortest 16% of all pregnancies are shorter than 250 days and the longest 16% of all pregnancies are longer than 282 days.

Use this reasoning with the middle 95% to determine how short the shortest 2.5% of all pregnancies are.

The shortest 2.5% of all pregnancies are less than ____________ days.

c) You should be able to do this on your own if you understand part b.

The longest 2.5% of all pregnancies are at least ______________ days.

COMPLETE SOLUTIONS

Exercise 2.2

a) A complete solution is given in the guided solution.

b) The area of interest is the shaded region indicated in the guided solution. This rectangular region has area = length of base × height = 0.6 × 1 = 0.6.

c) The area of interest is the shaded region indicated below.

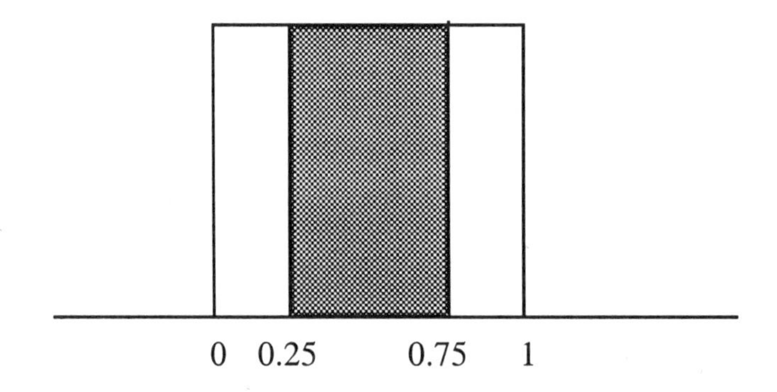

Note that the base has length 0.75 – 0.25 = 0.5. This rectangular region therefore has area = length of base × height = 0.5 × 1 = 0.5.

Exercise 2.12

a) A complete solution is given in the guided solution.

b) According to part a, 16% of soldiers have head circumferences larger than 23.9 inches. The remaining 84% must have head circumferences less than or equal to 23.9 inches. Therefore, a head circumference of 23.9 inches would be the 84th percentile.

c) Referring to the figure above part a in the guided solutions, we see that 68% of soldiers have head circumferences from 21.7 inches to 23.9 inches.

Exercise 2.13

a) The middle 95% lie between the mean minus two standard deviations and the mean plus two standard deviations. Because the mean is 266 days and the standard deviation is 16 days, the middle 95% are as in the figure below.

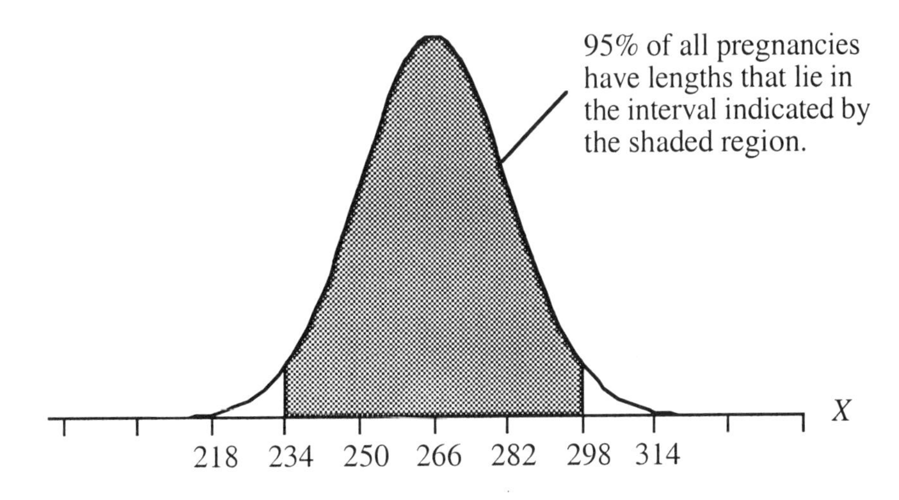

We see from the figure that the middle 95% of the lengths of all pregnancies have lengths from 234 to 298 days.

b) Refer to the figure in part a above. If the shaded region gives the middle 95% of the area, then the two unshaded regions must account for the remaining 5%. Because the normal curve is symmetric, each of the two unshaded regions must have the same area and each must account for half of the remaining 5%. Hence, each of the unshaded regions accounts for 2.5% of the area. The leftmost of these regions accounts for the shortest 2.5% of all pregnancies. We conclude:

The shortest 2.5% of all pregnancies are less than ____234____ days.

c) Refer to the figure in part a above. If the shaded region gives the middle 95% of the area, then the two unshaded regions must account for the remaining 5%. Because the normal curve is symmetric, each of the two unshaded regions must have the same area and each must account for half of the remaining 5%. Hence each of the unshaded regions accounts for 2.5% of the area. The rightmost of these regions accounts for the longest 2.5% of all pregnancies. We conclude:

The longest 2.5% of all pregnancies are at least ____298____ days.

SECTION 2.2

OVERVIEW

Areas under any normal curve can be found easily if variables (x) are first standardized by subtracting the mean (μ) from each value and dividing the result by the standard deviation (σ). This **standardized value** is called the ***z*-score**.

$$z = \frac{x - \mu}{\sigma}$$

If data whose distribution can be described by a normal curve are standardized (all values replaced by their z-scores), the distribution of these standardized values is described by the **standard normal distribution**. Areas under standard normal curves are easily computed by using a **standard normal table** such as the one found in Table A in the inside front cover of the textbook.

The standard normal curve is very useful for finding the proportion of observations in an interval when dealing with any normal distribution. Here are some hints about solving these problems:

1. State the problem.

2. Draw a picture of the problem. It will help to remind you of the area for which you are looking.

3. Standardize the observations.

4. Using Table A from the textbook, find the area you need.

HINT: The normal curve has a total area of 1. The normal curve is also symmetric so areas (proportions) such as those shown below are equal.

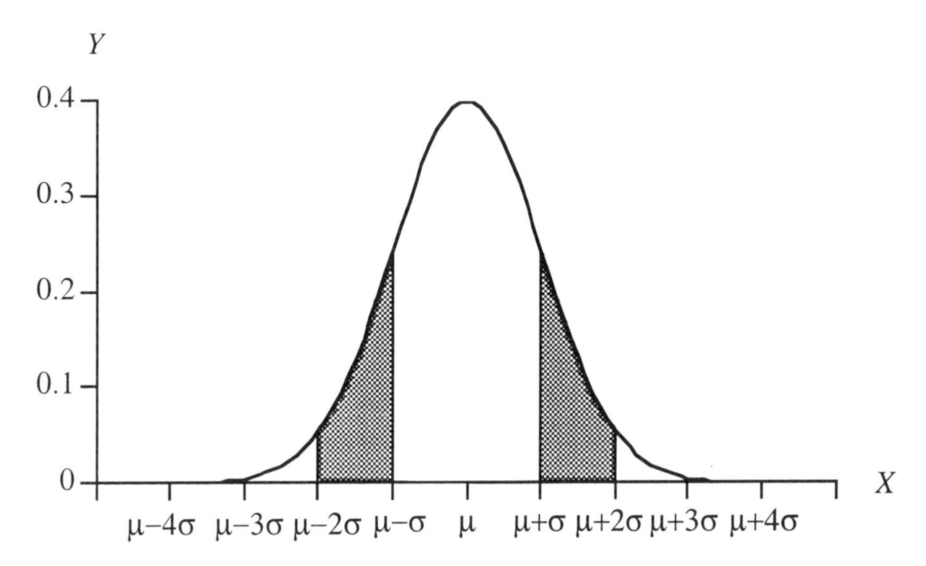

One also can perform "backward" or "reverse" normal calculations. In particular, if you are given a proportion and want to find the value corresponding to the proportion, the following steps are helpful:

1. State the problem.
2. Using Table A from the textbook, find the entry in the *body* of the table corresponding to the proportion you are given.
3. Unstandardize

If we know the distribution of data is described by a normal curve, we can make statements about what values are likely and unlikely, without actually observing the individual values. How do we decide if it is plausible to assume that data come from a normal distribution? One can examine a histogram or stem-and-leaf plot to see whether it is bell-shaped. Another method for determining whether the distribution of data is described by a normal curve is to construct a **normal probability plot**. These are easily made using a TI-83 calculator or modern statistical computer software. If the distribution of data is described by a normal curve, the normal probability plot should look like a straight line.

GUIDED SOLUTIONS

Exercise 2.23

KEY CONCEPTS: Finding the value z (the quantile) corresponding to a given area under a standard normal density curve

The strategy used to solve this type of problem is the "reverse" of that used to find the area under a given portion of a normal curve. We begin by drawing a picture of what we know; we know the area, but not z. For areas corresponding to those given in Table A, we have a situation like the following.

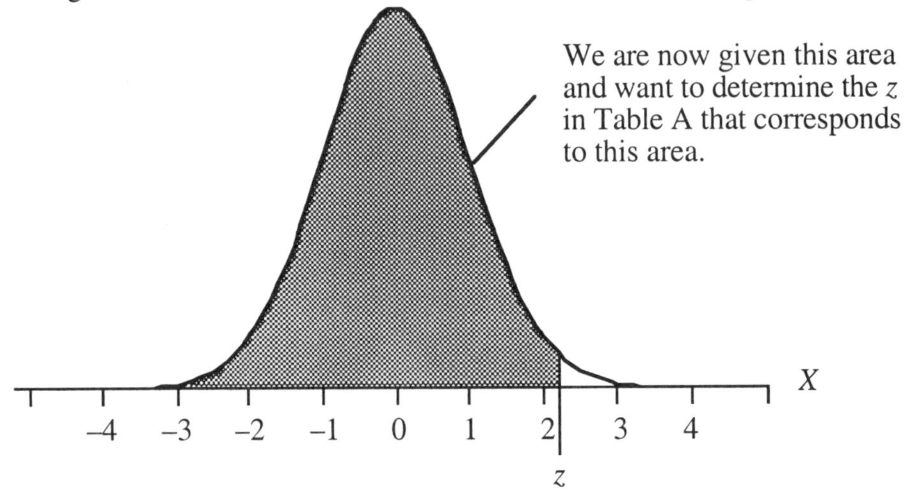

To determine z, we find the given value of the area in the body of Table A (or the entry in Table A closest to the given value of the area). Next, we look in the left margin of the table and across the top of the table to determine the value of z that corresponds to this area.

If we are given a more complicated area, we draw a picture and determine the area to the left of z from properties of the normal curve. Next, we determine z as described above. The approach is illustrated in the solutions below.

a) A picture of what we know is given below. Note that since the area given is less than 0.5, we know z must be to the left of 0 (recall that the area to the left of 0 under a standard normal curve is 0.5).

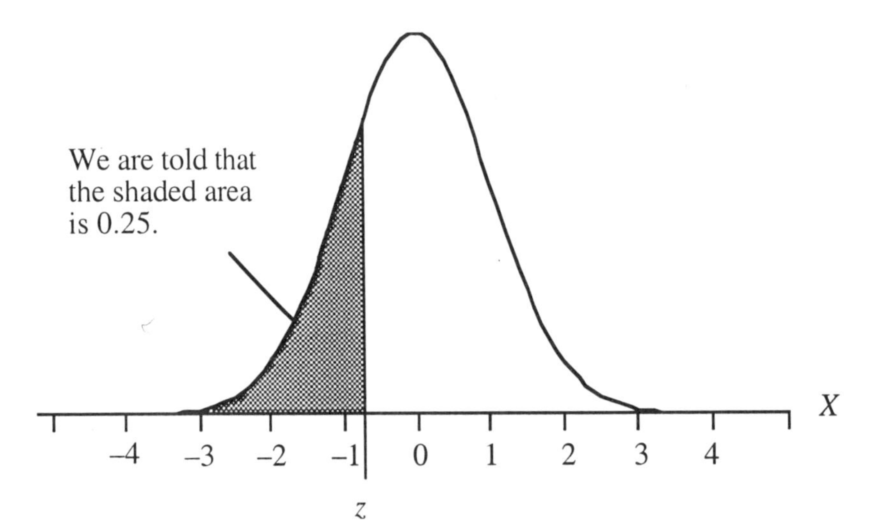

We now turn to Table A and find the entry closest to 0.25. This entry is 0.2514. Locating the z values in the left margin and top column corresponding to this entry, we see that the z that would give this area is –0.67.

b) Try this part on your own. Begin by sketching a normal curve and the area you are given on the curve. On which side of zero should z be located? Thinking about which side of zero a point lies on is a good way to make sure your answer makes sense.

Exercise 2.28

KEY CONCEPTS: Computing areas under a standard normal density curve

Recall that the proportion of observations from a standard normal distribution that are less than a given value z is equal to the area under the standard normal curve to the left of z. Table A gives these areas which are illustrated in the figure below.

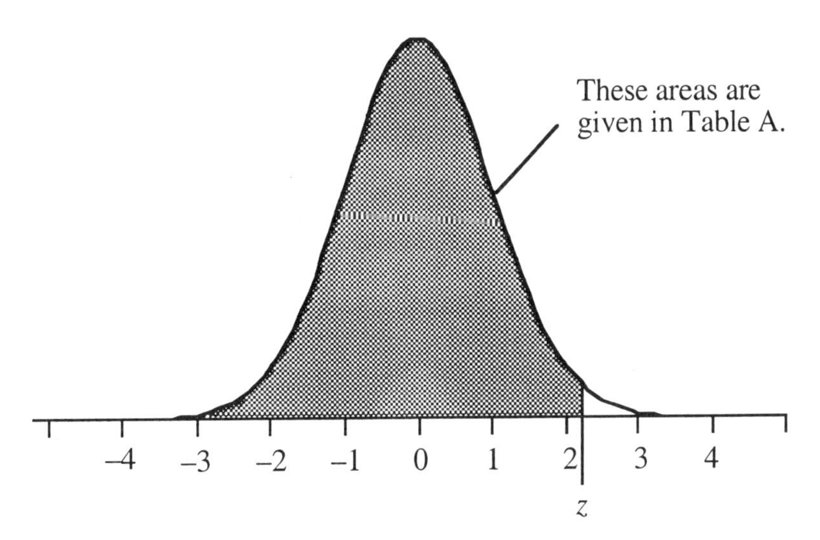

In answering questions concerning the proportion of observations from a standard normal distribution that satisfy some relation, we find it helpful to first draw a picture of the area under a normal curve corresponding to the relation. We then try to visualize this area as a combination of areas of the form in the figure above, because such areas can be found in Table A. The entries in Table A are then combined to give the area corresponding to the relation of interest.

This approach is illustrated in the solutions that follow.

a) To get you started, we will work through a complete solution. A picture of the desired area is given below.

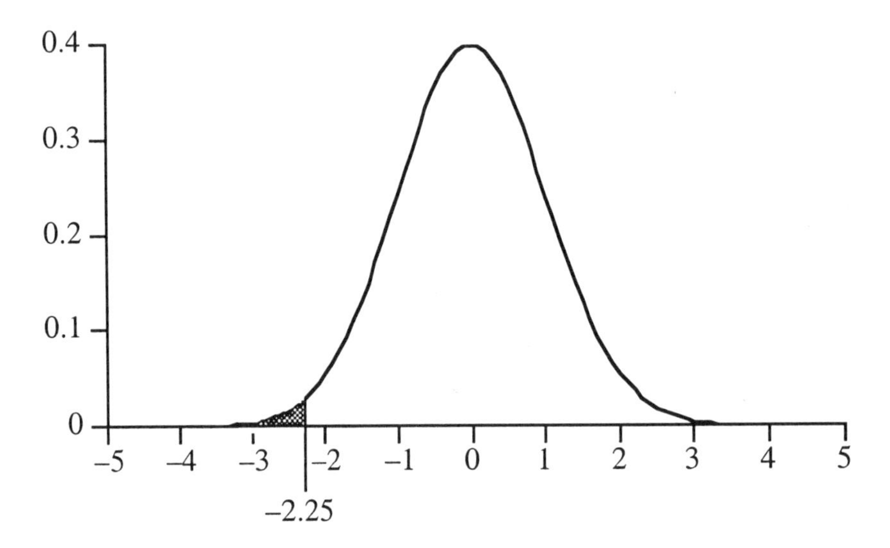

This is exactly the type of area that is given in Table A. We simply find the row labeled –2.2 along the left margin of the table, locate the column labeled .05 across the top of the table, and read the entry in the intersection of this row and column. This entry is 0.0122, which is the proportion of observations from a standard normal distribution that satisfies $z < -2.25$.

b) Shade the desired area in the figure below.

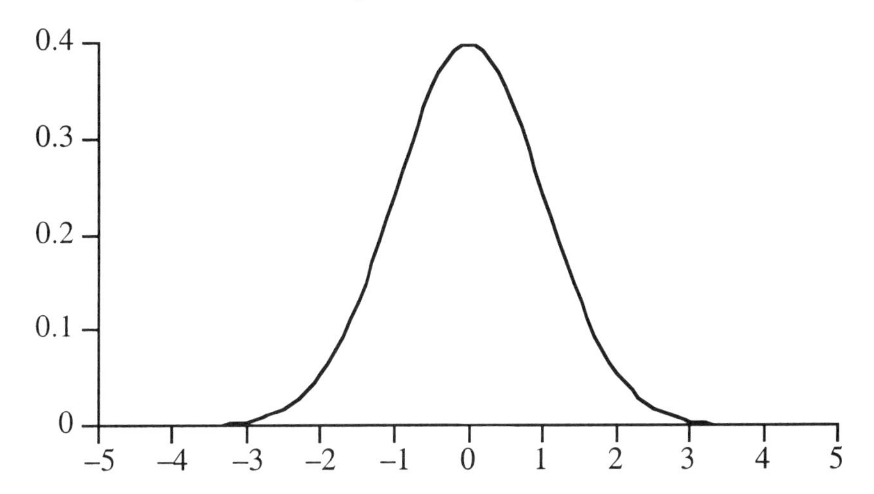

Remembering that the area under the whole curve is 1, how would you modify your answer from part a?

area =

c) Try solving this part on your own. To begin, draw a picture of a normal curve and shade the region.

Next, use the same line of reasoning as in part b to determine the area of your shaded region. Remember, you want to try to visualize your shaded region as a combination of areas of the form in given in Table A.

d) To test yourself, try this part on your own. It is a bit more complicated than the previous parts, but the same approach will work. Draw a picture and then try to express the desired area as the difference of two regions for which the areas can be found directly in Table A.

Exercise 2.34

KEY CONCEPTS: Computing areas under a standard normal density curve

You are asked to verify the answers to exercise 2.28 (the previous exercise discussed above).

a) To find the proportion of observations from a standard normal distribution below –2.25, use the following keystrokes.

2nd/DISTR/2:normalcdf(

Fill in the parentheses as (–1E99, –2.25) followed by hitting the ENTER key. Remember to use the keystrokes (–)/1/2nd EE/99 to produce –1E99.

You should get .0122244334. This agrees with the answer to part a of exercise 2.28.

b) This time you will want to use the keystrokes

2nd/DISTR/2:normalcdf(

and fill in the parentheses as (–2.25, 1E99). What do you get?

ANS = ____________________

c) Try this problem on your own.

ANS = ___________________

d) Do this one on your own.

ANS = ___________________

Exercise 2.41

KEY CONCEPTS: Computing proportions for an arbitrary normal density curve

The steps for finding proportions for arbitrary normal curves are:

1. State the problem.

2. Draw a picture of the problem. It will help remind you of the area for which you are looking.

3. Standardize the observations.

4. Using Table A from the textbook, find the area you need.

We sketch these steps for this problem. You should fill in the details.

1. The normal distribution for the reference population has mean $\mu = 100$ and standard deviation $\sigma = 15$. Do we want the proportion with scores greater than 135 or the proportion with scores less than 135?

2. Shade the area that corresponds to the desired proportion in the graph.

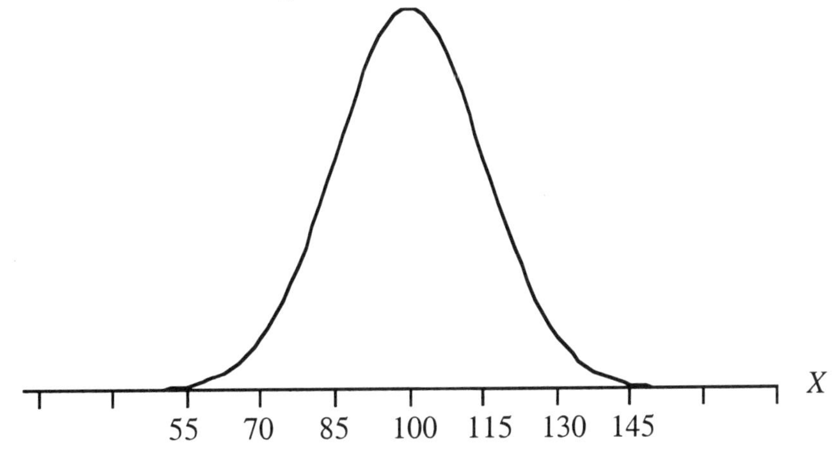

3. Standardize the value 135 by completing the following.

$$z = \frac{135 - \mu}{\sigma} =$$

4. Use Table A to compute the area corresponding to the shaded region in step 2 under a standard normal curve. This area is the desired proportion.

To complete the problem, multiply 1300 by this proportion to estimate how many sixth graders in the school district are gifted.

Exercise 2.43

KEY CONCEPTS: Finding the value x corresponding to a given proportion or area under an arbitrary normal density curve

To solve this problem, we must use an approach that is the reverse of the one used in exercise 2.41. Conceptually, we think of having standardized the problem. We then find the value z for the standard normal distribution that satisfies the stated condition, i.e., has the desired area. Next, we must "unstandardize" this z value by multiplying by the standard deviation and adding the mean to the result. This unstandardized value x is the desired result.

Example 2.8 in the textbook summarizes the steps needed for such a "backward" normal calculation. We follow these steps.

State the problem. We first want to find the head size x_1 with area 0.05 (corresponding to 5%) to the left under the normal curve with mean $\mu = 22.8$ inches and standard deviation $\sigma = 1.1$ inches. This will tell us the head size below which are the smallest 5% of head sizes. We also want to find the head size x_2 with area 0.05 (corresponding to 5%) to the right under the normal curve with mean $\mu = 22.8$ inches and standard deviation $\sigma = 1.1$ inches. This will tell us the head size above which are the largest 5% of head sizes.

The figure below states this question in graphical form.

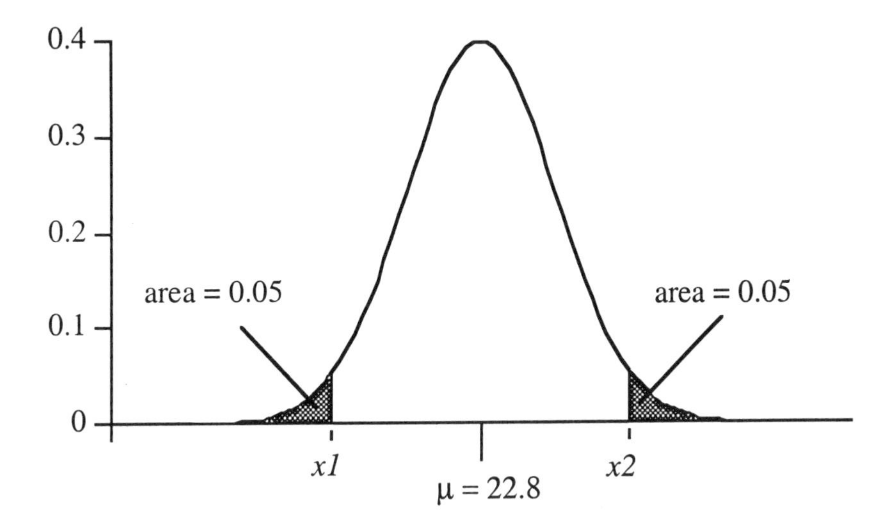

We need to express these areas as areas to the left of x_1 and x_2. What are these areas?

Use the table. Look in Table A for the entry closest to these two areas. For example, the area to the left of x_1 is 0.05. The entry in Table A with this area is about $z_1 = -1.65$. Thus $z_1 = -1.65$ is the standardized value with area 0.05 to its left. What is the area to the left of x_2, and what is the standardized value z_2 with this area to its left?

$z_2 =$

Unstandardize. To transform the solution from z_1 and z_2 back to the original x scale, we use the "unstandardizing" formula

$$x = \mu + z\sigma.$$

Applying this formula to z_1 yields

$$x_1 = 22.8 + (z_1 \times 1.1) = 22.8 + (-1.65 \times 1.1) = 22.8 - 1.815 = 20.985 \text{ in.}$$

Next, you find x_2.

$x_2 =$

State your conclusions.

Exercise 2.46

KEY CONCEPTS: Normal probability plots

Use either statistical software or your calculator to make the plot. If you are using your calculator, to make the normal probability plot for the normal corn, first enter the data in problem 1.50 for normal corn as a list (say, L1). Next, use the sequence 2nd/STAT PLOT. Make sure all plots are off except for Plot 2. Define Plot 2 like this:

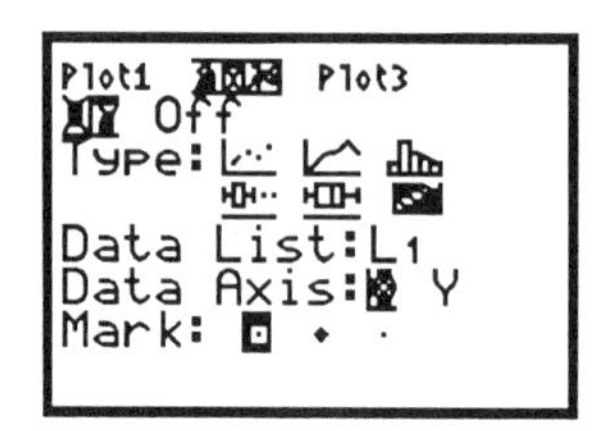

Then use the key sequence ZOOM/9:ZoomStat to display your plot.

Repeat this for the new corn. What do you observe? Is the use of $\bar{x}$ and s justified?

COMPLETE SOLUTIONS

Exercise 2.23

a) A complete solution was given in the guided solution.

b) A picture of what we know is given below. Note that because the area to the right of 0 under a standard normal curve is 0.5, we know that z must be located to the right of 0.

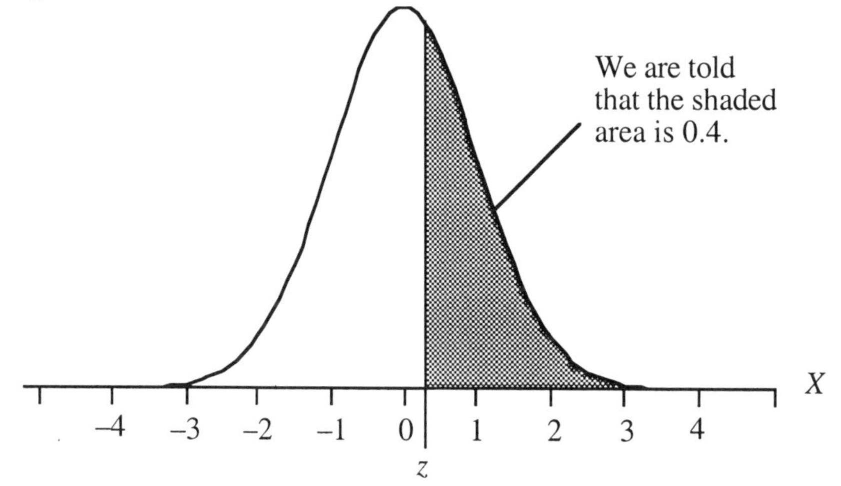

The shaded area is not of the form used in Table A. However, we note that the unshaded area to the left of z is of the correct form. Because the total area under a normal curve is 1, this unshaded area must be $1 - 0.4 = 0.6$. Hence z has the property that the area to the left of z must be 0.6. We locate the entry in Table A closest to 0.6. This entry is 0.5987. The z corresponding to this entry is 0.25.

Exercise 2.28

a) A complete solution was provided in the guided solutions.

b) The desired area (see figure below) is not of the form for which Table A can be used directly. However, the unshaded area to the left of –2.25 is of the form needed for Table A. In fact, we found the area of the unshaded portion in part a. We notice that the shaded area can be visualized as what is left after deleting the unshaded area from the total area under the normal curve.

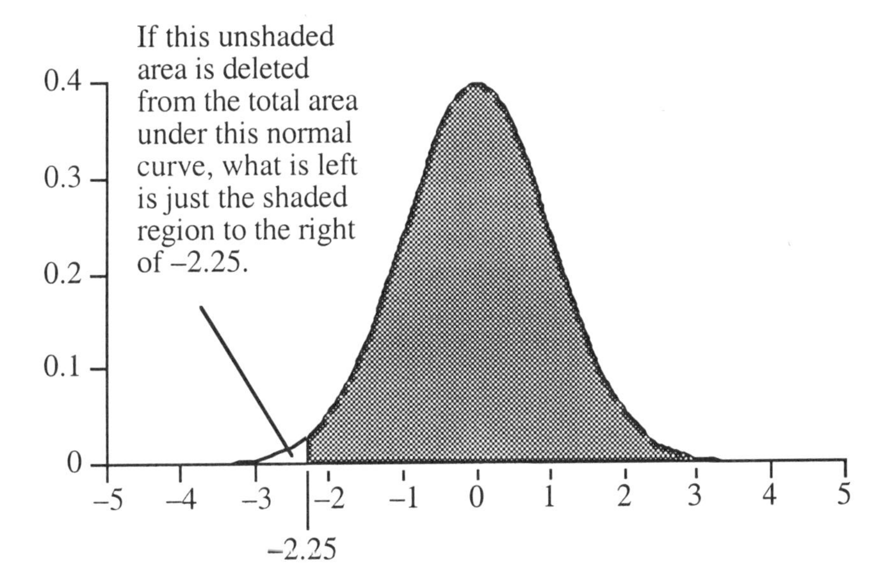

Since the total area under a normal curve is 1, we have

$$\begin{aligned}\text{shaded area} &= \text{total area under normal curve} - \text{area of unshaded portion}\\ &= 1 - 0.0122 = 0.9878.\end{aligned}$$

Thus, the desired proportion is 0.9878.

c) The desired area is indicated in the figure below.

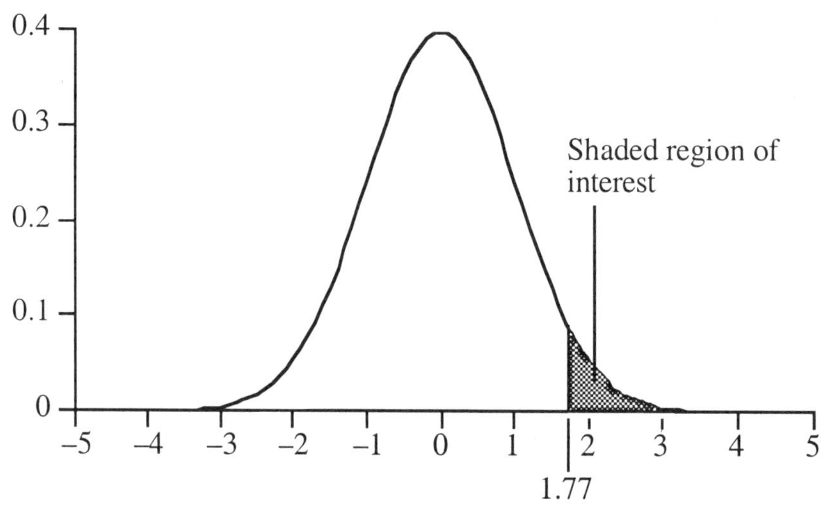

This is just like part b. The unshaded area to the left of 1.77 can be found in Table A and is 0.9616. Thus,

shaded area = total area under normal curve – area of unshaded portion
= 1 – 0.9616 = 0.0384.

This is the desired proportion.

d) We begin with a picture of the desired area.

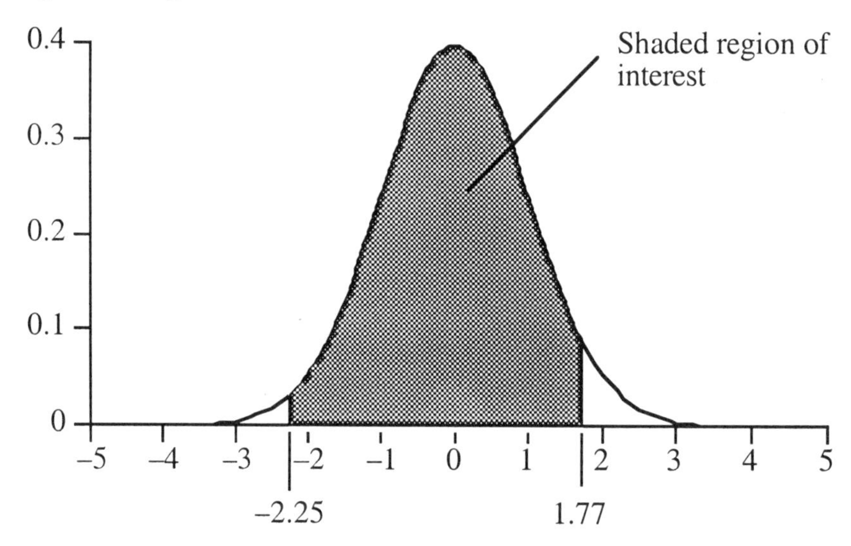

While the shaded region is a bit more complicated than in the previous parts, the same strategy still works. We note that the shaded region is obtained by removing the area to the left of –2.25 from all of the area to the left of 1.77.

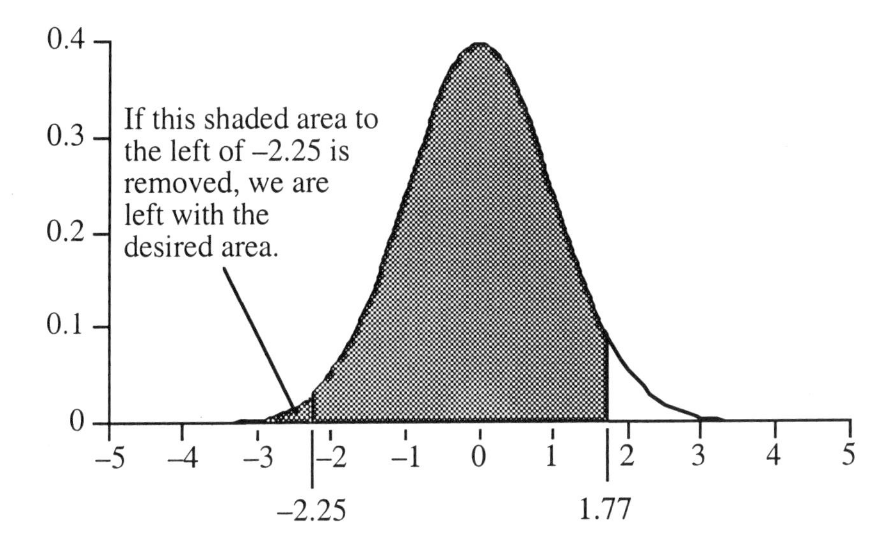

The area to the left of –2.25 is found to be 0.0122 in Table A. The area to the left of 1.77 is found to be 0.9616 in Table A. Thus, the

shaded area = area to left of 1.77 – area to left of –2.25
= 0.9616 – 0.0122
= 0.9494.

This is the desired proportion.

Exercise 2.34

a) A complete solution is provided in the guided solutions.

b) Using the keystrokes given, you should get the following:

ANS = .9877755666.

This agrees with the answer to part b of exercise 2.28.

c) The keystrokes are 2nd/DISTR/2:normalcdf(. The parentheses should be filled in as (1.77, 1E99). You should get

ANS = .0383635226.

This agrees with the answer to part c of exercise 2.28.

d) The keystrokes are 2nd/DISTR/2:normalcdf(. The parentheses should be filled in as (–2.25, 1.77). You should get

ANS = .949412044.

This agrees with the answer to part d of exercise 2.28.

Exercise 2.41

1. We want the proportion with scores greater than 135.

2. The desired area is shaded in the following graph.

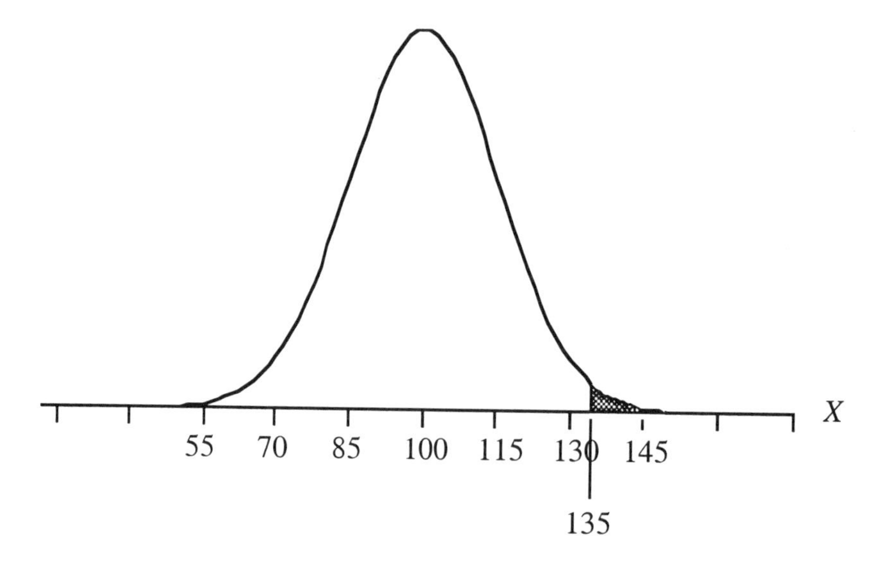

3. The standardized value is

$$z = \frac{135-\mu}{\sigma} = \frac{135-100}{15} = \frac{35}{15} = 2.33.$$

4. Table A tells us that the area under a standard normal curve to the left of 2.33 is 0.9901. We want the area to the right of 2.33, so we subtract 0.9901 from 1 to get 0.0099. Thus,

the proportion with scores greater than 135 = 0.0099

We would estimate that 0.0099 × 1300 = 12.87, or approximately 13 sixth graders in the school district are gifted.

Exercise 2.43

From the figure we see that the area to the left of x_1 is 0.05 and to the left of x_2 is 0.95. The corresponding standardized values are

$$z_1 = -1.65$$

$$z_2 = +1.65$$

The unstandardized values are

$$x_1 = 22.8 + (z_1 \times 1.1) = 22.8 + (-1.65 \times 1.1) = 22.8 - 1.815 = 20.985 \text{ inches}$$

and

$$x_2 = 22.8 + (z_2 \times 1.1) = 22.8 + (1.65 \times 1.1) = 22.8 + 1.815 = 24.615 \text{ inches}$$

We conclude that soldiers in the smallest 5% have head sizes below 20.985 inches and those in the largest 5% have head sizes above 24.615 inches. Thus, soldiers with head sizes below 20.985 or above 24.615 inches get custom-made helmets.

Exercise 2.46

Here are the TI-83 normal probability plots:

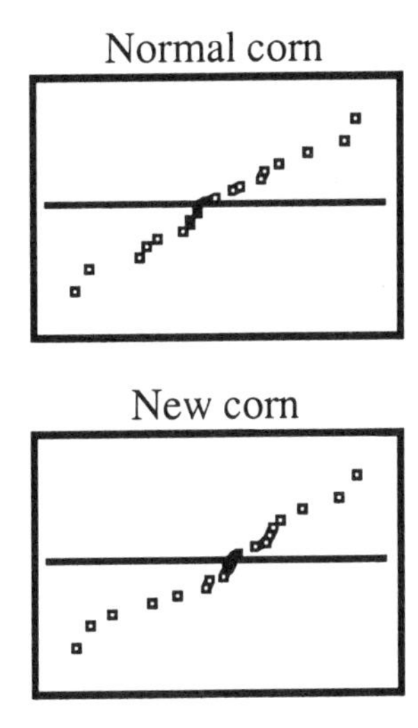

Neither plot is perfectly linear, but neither suggests any severe departures from normality. The use of $\bar{x}$ and s is not unreasonable.

CHAPTER 3

EXAMINING RELATIONSHIPS

SECTION 3.1

OVERVIEW

Chapter 1 provides the tools to explore several types of variables one by one, but in most instances the data of interest are a collection of variables that may exhibit relationships among themselves. Typically, these relationships are more interesting than the behavior of the variables individually. The first tool we consider for examining the relationship between variables is the **scatterplot**. Scatterplots show us two quantitative variables at a time, such as the weight of a car and its miles per gallon (MPG). Using colors or different symbols, we can add information to the plot about a third variable that is categorical in nature. For example, if in our plot we wanted to distinguish between cars with manual or automatic transmissions, we might use a circle to plot the cars with manual transmissions and a cross to plot the cars with automatic transmissions.

When drawing a scatterplot, we need to pick one variable to be on the horizontal axis and the other to be on the vertical axis. When there is a **response variable** and an **explanatory variable**, the explanatory variable is always placed on the horizontal axis. In cases where there is no explanatory-response variable distinction, either variable can go on the horizontal axis. After drawing the scatterplot by hand or using a computer, the scatterplot should be examined for an **overall pattern** that may tell us about any relationship between the variables and for **deviations** from it. You should be looking for the **direction**, **form**, and **strength** of the overall pattern. In terms of direction, **positive association** occurs when both variables take on high values together, while **negative association** occurs if one variable takes high values when the other takes on low values. In many cases, when an association is present, the variables appear to have a **linear relationship**. The plotted values seem to follow a line. If the line slopes up to the right, the association is positive; if the line slopes down to the right, the association is negative. As always, look for **outliers**. The outlier may be far away in terms of the horizontal variable or the vertical variable or far away from the overall pattern of the relationship.

GUIDED SOLUTIONS

Exercise 3.3

KEY CONCEPTS: Explanatory and response variables, categorical and quantitative variables

When examining the relationship between two variables, if you want to show that one of the variables can be used to explain variation in the other, remember that the **response variable** measures the outcome of the study, while the **explanatory variable** explains or is associated with changes in the response variable. When you just want to explore the relationship between two variables, such as score on the math and verbal SAT, then the explanatory-response variable distinction is not important.

If you do not recall the distinction between a categorical and a quantitative variable, refer to the Introduction to Chapter 1.

In this problem, the response variable is how long a patient lived after treatment, because it is the outcome of the study. Is this variable categorical or quantitative?

Next, it is your turn.

explanatory variable =

Is this variable categorical or quantitative?

Exercise 3.9

KEY CONCEPTS: Drawing and interpreting a scatterplot, adding a categorical variable to a scatterplot

a) When drawing a scatterplot, we first need to pick one variable (the explanatory variable) to be on the horizontal axis and the other (the response) to be on the vertical axis. In this data set, we are interested in the "effect" of lean body mass on metabolic rate. So lean body mass is the explanatory variable and metabolic rate is the response variable. We have labeled the axes appropriately in the plot in the following figure. Try plotting the points in the figure. Although you will generally draw scatterplots using a calculator or computer, drawing a small example like this by hand makes sure that you understand what the points represent.

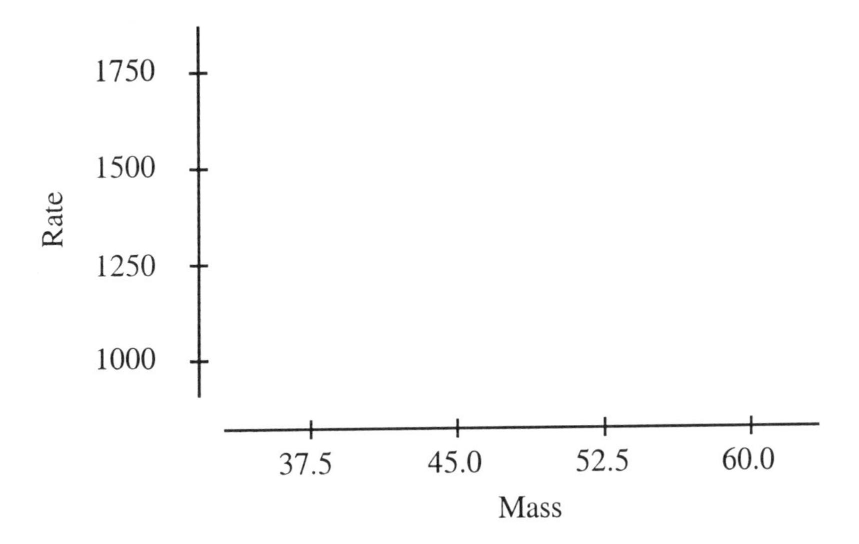

b) Here are some guidelines for examining scatterplots: Do the data show any association? **Positive association** is present when both variables take on high values together and both take on low values together. **Negative association** is present when high values of one variable are accompanied with low values of the other. If the plotted values seem to follow a line, the variables may have a **linear relationship**. If the line slopes up to the right, the association is positive. If the line seems to slope down to the right, the association is negative. Are there any **clusters** of data? Clusters are distinct groups of observations. As always, look for outliers. An **outlier** may be far away in terms of the horizontal variable or the vertical variable, or far away from the overall pattern of the relationship.

For our plot, is the association positive or negative? Do females with higher lean body mass tend to have higher or lower metabolic rates?

What is the form of the relationship? Do the points in the plot tend to follow a straight-line pattern? A curved pattern? Are distinct clusters present?

How strong is the relationship?

c) The scatterplot for the females is presented here. Add the data for the males to this plot using a different color or plotting symbol.

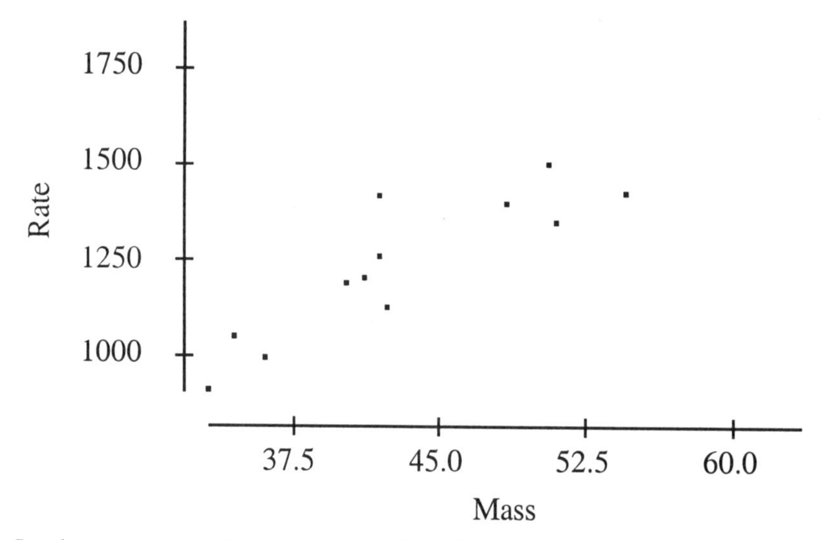

Is the pattern of the relationship for the men similar to that for the female subjects? If not, how does it differ?

How do the male subjects as a group differ from the female subjects as a group?

Exercise 3.12

KEY CONCEPTS: Drawing and interpreting a scatterplot, outliers

a) If you wish to make the scatterplot by hand, draw your scatterplot on the axes that follow. Otherwise, use your calculator or computer to create the plot. Note that because we wish to predict guessed calories from correct calories, correct calories is the predictor variable while guessed calories is the response.

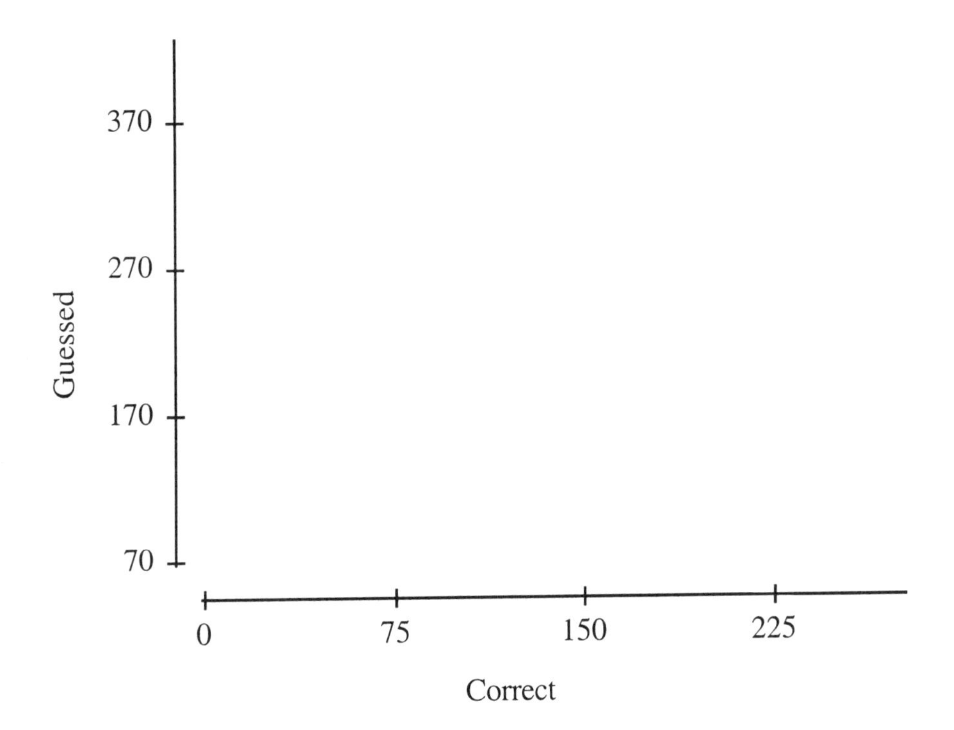

b) How would you describe the relationship in your plot? Is it roughly linear? Curved? No clear relation? Is the association positive (high values of the two variables tend to occur together), negative (high values of one variable tend to occur with low values of the other), or is there no clear pattern? Are there any outliers (points that deviate from the overall pattern)?

Exercise 3.13

KEY CONCEPTS: Drawing and interpreting scatterplots

a) For these data, we do not envision one of the variables as explaining the other. All we are interested in is investigating the association between the two variables. We are free to arbitrarily designate one of the variables as the explanatory variable and the other as the response. We choose the variable Femur as the explanatory variable, plotting it on the horizontal axis, and choose Humerus to play the role of the response, plotting it on the vertical axis. If you wish to make the scatterplot by hand, draw your scatterplot on the axes that follow. Otherwise, use your calculator or computer to create the plot. Because there are only five points, it might be easier to draw the plot yourself rather than use a calculator or computer.

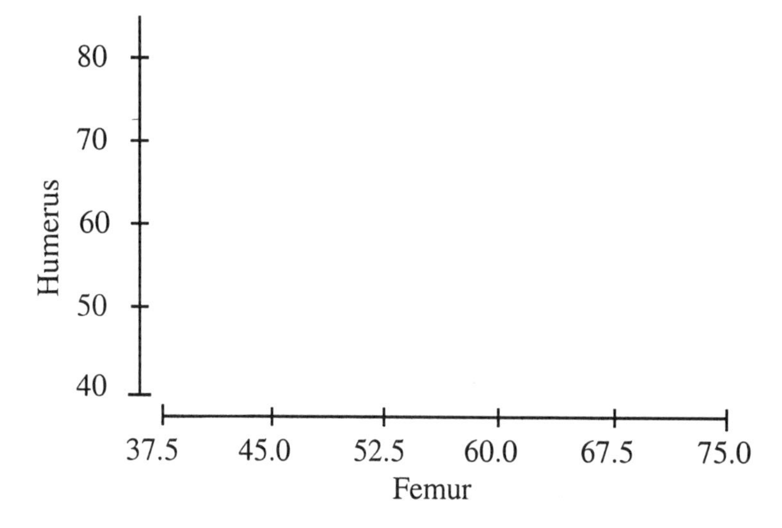

Do any of the points appear to depart from a straight line? What does this imply about whether all five specimens came from the same species?

COMPLETE SOLUTIONS

Exercise 3.3

As stated in the guided solution, the response variable is how long a patient lived after treatment. Because it is a measure of time, it is a quantitative variable.

The explanatory variable is the treatment (the removal of the breast versus the removal of only the tumor and nearby lymph nodes) a patient received. The treatments attempt to explain the response (how long after treatment a patient lived). The explanatory variable is categorical because it classifies the individuals into one of two treatment categories.

Exercise 3.9

a) The plot follows.

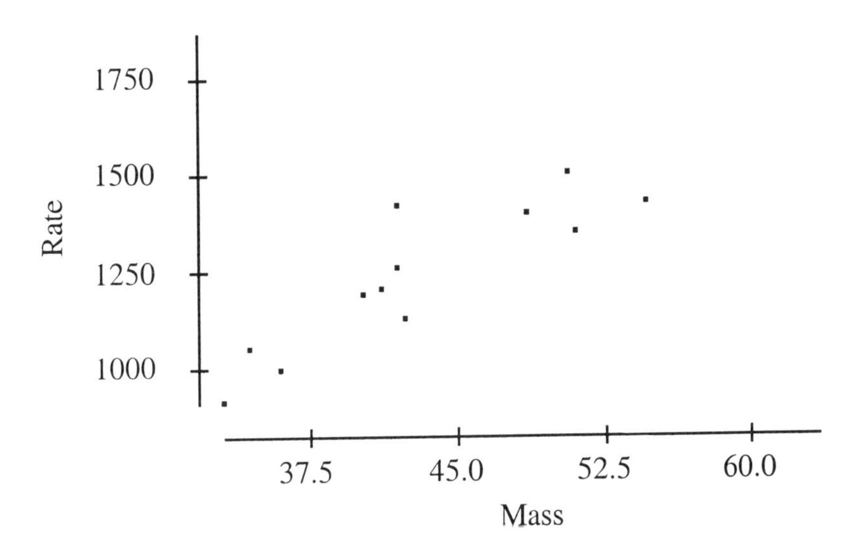

b) As lean body mass increases, or as you move from left to right across the horizontal axis in the scatterplot, the points in the plot tend to rise. This indicates that the association between the variables is positive. The form of the relationship appears to be linear because a straight line seems to be a reasonable approximation to the overall trend in the plot. The relationship is not perfect, but it appears to be moderately strong.

c) We add the men to the plot. Men are indicated by the x's.

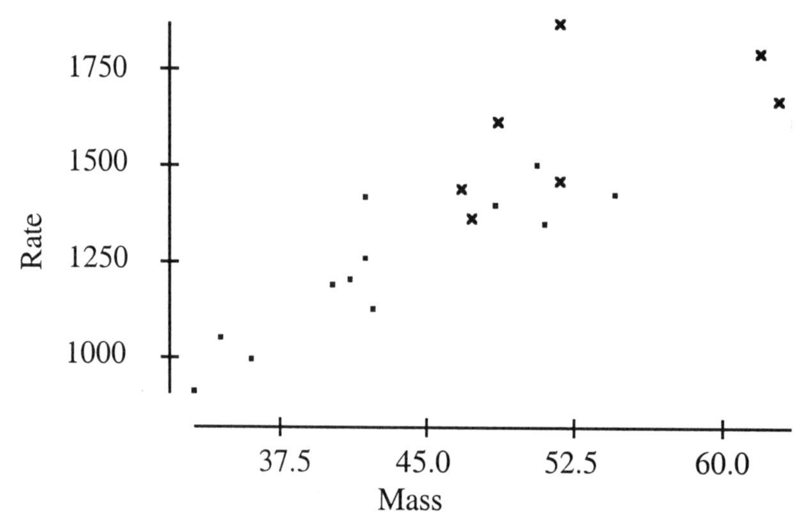

The male subjects (the x's) also show a positive association that might be described as linear. The association does not appear so strong as for the women and the slope of the linear relation may be a bit flatter. We also notice that the men are clustered in the upper right of the plot. This is not surprising, because men tend to be larger than women.

Exercise 3.12

a, b) Here is the scatterplot.

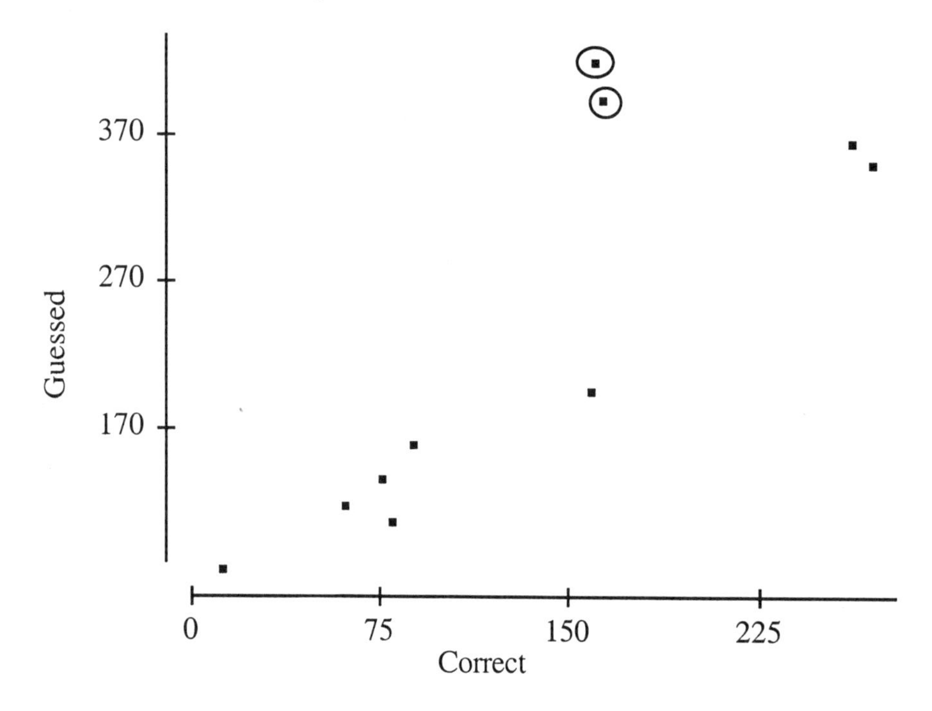

The relationship appears to be linear and the association is positive, because the trend is for the guessed and correct values to be both small or both large together. The two circled points are outliers.

Exercise 3.13

a) The scatterplot follows.

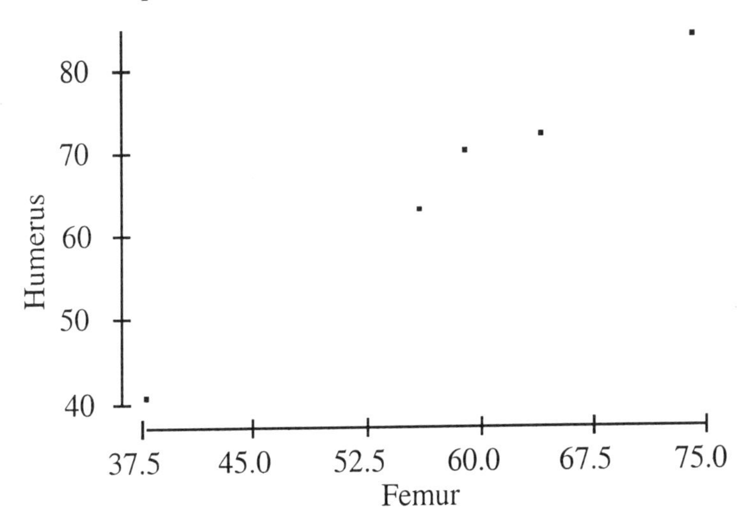

One of the points (the third point, with Femur = 59 and Humerus = 70) appears to differ a bit from the others. If this point is ignored, the other four appear to lie almost exactly on a straight line. While the difference does not appear to be dramatic, this point might come from a different species than the others. The evidence does not appear overwhelming, however.

SECTION 3.2

OVERVIEW

Scatterplots provide a visual tool for looking at the relationship between two variables. Unfortunately, our eyes are not good tools for judging the strength of the relationship. Changes in the scale or the amount of white space in the graph can easily change our judgment of the strength of the relationship. **Correlation** is a numerical measure we use to show the strength of **linear association**.

The correlation can be calculated using the formula

$$r = \frac{1}{n-1}\sum\left(\frac{x_i - \bar{x}}{s_x}\right)\left(\frac{y_i - \bar{y}}{s_y}\right)$$

where $\bar{x}$ and $\bar{y}$ are the respective means for the two variables X and Y, and s_X and s_Y are their respective standard deviations. In practice, you will probably be computing the value of r using computer software or a calculator that finds r from entering the values of the x's and y's. When computing a correlation coefficient, there is no need to distinguish between the explanatory and response variables, even in cases where this distinction exists. The value of r does not change if we switch x and y. The value of r also does not change if we change the unit of measurement for either variable.

When r is positive there is a positive linear association between the variables, and when r is negative there is a negative linear association. The value of r is always between 1 and –1. Values close to 1 or –1 show a strong linear association, while values near 0 show a weak linear association. As with means and standard deviations, the value of r is strongly affected by outliers. Their presence can make the correlation much different than it might be with the outlier removed. Finally, remember that the correlation is a measure of straight line association. There are many other types of association between two variables, but these patterns will not be captured by the correlation coefficient.

GUIDED SOLUTIONS

Exercise 3.19

KEY CONCEPTS: Computing the correlation coefficient

a) Let x denote the Femur measurements and y the Humerus measurements. We find the means and standard deviations to be

$$\bar{x} = 58.20, \quad s_X = 13.1985$$

$$\bar{y} = 66.00, \quad s_Y = 15.8902$$

Calculations by hand are best done systematically, such as in the following table. The second and fourth columns are the standardized values for x and y. We have provided the table entries for the first two x, y values. See whether you can complete the remaining entries.

x	$\left(\frac{x-\bar{x}}{s_x}\right)$	y	$\left(\frac{y-\bar{y}}{s_y}\right)$	$\left(\frac{x-\bar{x}}{s_x}\right)\left(\frac{y-\bar{y}}{s_y}\right)$
38	–1.5305	41	–1.5733	2.4079
56	–0.1667	63	–0.1888	0.0315
59		70		
64		72		
74		84		

Now sum up the values in the last column and divide by $n-1$ to compute r.

$r =$

b) Follow the instructions given in the statement of the problem in the textbook. Our calculator gives a correlation of 0.9941485714. You should get a similar result. Although we have displayed the result to the number of decimal places given by the calculator, normally we would round off to two or three places. The result agrees with the result in part a.

Exercise 3.25

KEY CONCEPTS: Interpreting and computing the correlation coefficient

a) If you wish to make the scatterplot by hand, draw your scatterplot on the axes that follow. Otherwise, use your calculator or computer to create the plot. For simplicity, consider using the symbols in the key.

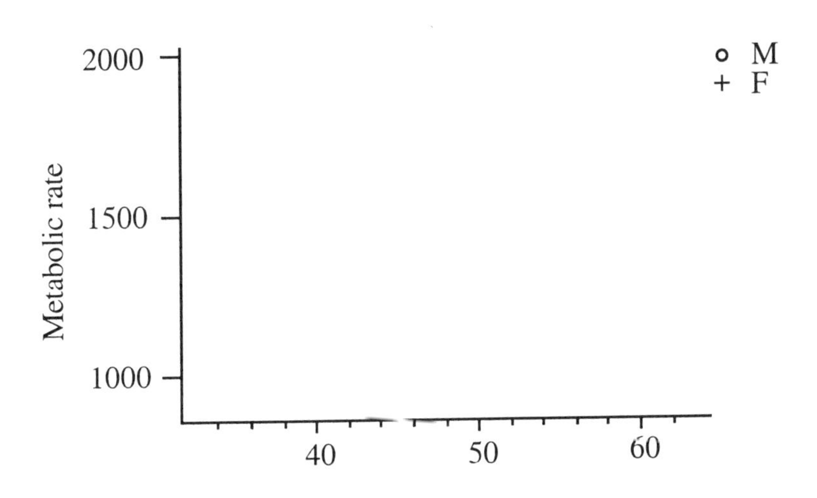

Should the sign of the correlation coefficient be the same for men and women? Is either relationship "stronger"? Are there outliers in either group that might raise or lower the value of the correlation coefficient?

b) Use your calculator (or computer if you are using statistical software) to compute the correlation coefficient. If you are using a TI-83, the steps are described in part b of problem 3.19 in your textbook. We reproduce the steps here for convenience.

Enter the data into your calculator in lists L_1 ans L_2 (explanatory variable in L_1 and response variable in L_2). Then key in this sequence: 2nd/CATALOG/x^{-1} (for the letter D)/(arrow down to select DiagnosticOn)/ENTER/ENTER. The calculator should say "Done." (Selecting DiagnosticOn is a one-time action.) Then key in this sequence: STAT/CALC/8:LinReg(a+bx)/ENTER. The calculator will display several lines, the last of which shows the correlation coefficient *r*.

Record your results.

Men's correlation coefficient =

Women's correlation coefficient =

c) This simply involves computing means. Record your results below.

Mean body mass for men =

Mean body mass for women =

This is a difficult question. If the relationship remained the same for all weights between 40 and 65 kilograms (the range of all the data), then it wouldn't matter. The fact that the men were heavier wouldn't influence the correlation. If the relationship were different for heavier people than lighter people, then the fact that men are heavier could affect the value of the correlation, because then men and women might have a different relationship between body mass and metabolic rate (with possibly differing strengths as measured by the correlation coefficient). What does the scatterplot suggest?

d) In general, what is the effect of changing the unit of measurement on the correlation coefficient?

Exercise 3.26

KEY CONCEPTS: Scatterplot, outliers, and correlation

a) If you wish to make the scatterplot by hand, draw your scatterplot on the axes that follow. Otherwise, use your calculator or computer to create the plot. Note that because we wish to predict guessed calories from correct calories, correct calories is the predictor variable while guessed calories is the response.

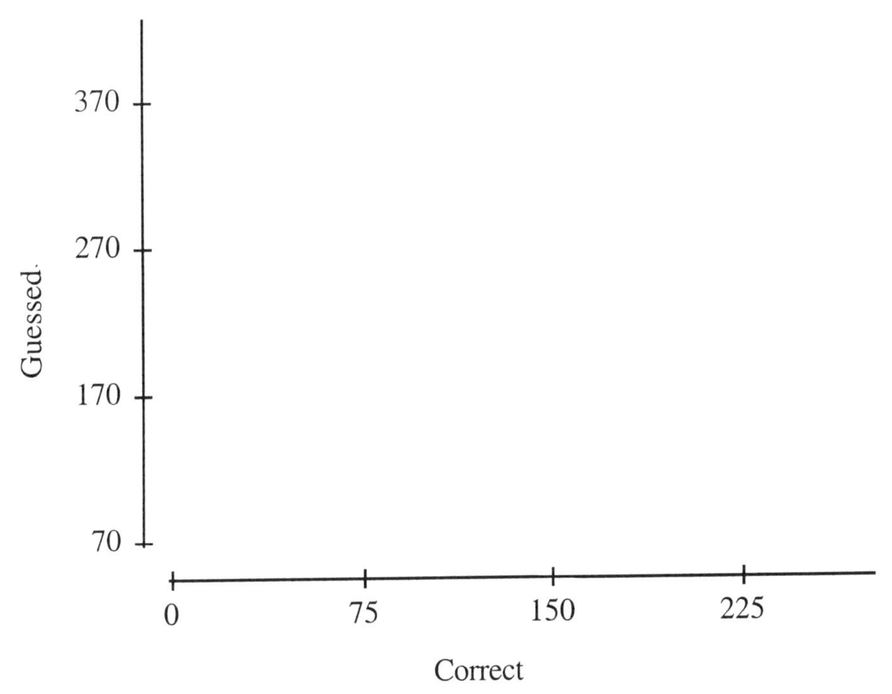

Use your calculator to compute r. See the guided solution for exercise 3.25 for the steps needed to compute r on the TI-83. Enter your answer in the space provided.

$r =$ ____________________________

Based on the scatterplot, why is this value reasonable?

b) Recall that the correlation is not affected by changing the units of measurement. How does this fact apply to the questions posed?

c) Circle the points in your plot in part a. Next, compute r for the other eight foods, leaving out these two points and record your answer in the space provided.

$r =$ ____________________

Why did r change as it did?

Exercise 3.28

KEY CONCEPTS: Interpreting the correlation coefficient

The problem is that a correlation close to zero and the quote "good researchers tend to be poor teachers, and vice versa" are not the same. What does a correlation close to zero mean? What would be true about the correlation if "good researchers tend to be poor teachers, and vice versa"?

COMPLETE SOLUTIONS

Exercise 3.19

a) The completed table follows.

x	$\left(\frac{x-\bar{x}}{s_x}\right)$	y	$\left(\frac{y-\bar{y}}{s_y}\right)$	$\left(\frac{x-\bar{x}}{s_x}\right)\left(\frac{y-\bar{y}}{s_y}\right)$
38	−1.5305	41	−1.5733	2.4079
56	−0.1667	63	−0.1888	0.0315
59	0.0606	70	0.2517	0.0153
64	0.4394	72	0.3776	0.1659
74	1.1971	84	1.1328	1.3561

The sum of the values in the last column is 3.9767. Thus, the correlation is

$$r = 3.9767/4 = 0.9942.$$

c) Our calculator gives a correlation of 0.9941485714. This agrees with the result in part a. Although we have displayed the correlation to the number of decimal places given by the calculator, normally we would round the result to only two or three places.

Exercise 3.25

a) The scatterplot follows.

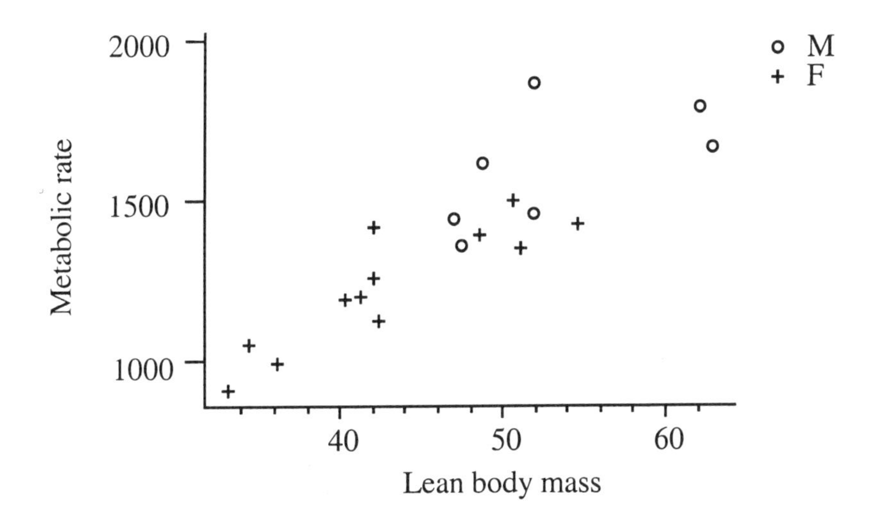

The relationship for the women seems to be more tightly clustered around a line. The men's observation of mass = 51.9 and rate = 1867 lowers the value of the correlation for the men.

b) You should try to use a calculator (or computer software) to compute the value of the correlation coefficient. If you do not have access to a calculator or computer package, or if you do not trust your calculations, the required "hand" computations are given below for the men. You can use them to double check your work if you made an error.

$$\bar{x} = 53.10, \quad s_x = 6.69$$

$$\bar{y} = 1600.00, \quad s_y = 189.2$$

We summarize the calculations for the correlation r in the following table

x	$\left(\frac{x-\bar{x}}{s_x}\right)$	y	$\left(\frac{y-\bar{y}}{s_y}\right)$	$\left(\frac{x-\bar{x}}{s_x}\right)\left(\frac{y-\bar{y}}{s_y}\right)$
62.0	1.33034	1792	1.01480	1.35003
62.9	1.46487	1666	0.34884	0.51100
47.4	–0.85202	1362	–1.25793	1.07178
48.7	–0.65770	1614	0.07400	–0.04867
51.9	–0.17937	1460	–0.73996	0.13273
51.9	–0.17937	1867	1.41121	–0.25313
46.9	–0.92676	1439	–0.85095	0.78862

The sum of the values in the last column is 3.5524. Thus, the correlation is

$$r = 3.5524/6 = 0.592$$

for the men. Thus,

Men's correlation coefficient = 0.592
(using a calculator we get 0.5920653037)

Likewise for the women we find

Women's correlation coefficient = 0.876
(using a calculator we get 0.8764526758)

c) In part b above, we see that

Mean body mass for men = 53.10.

If we carry out a similar computation for women, we find that

Mean body mass for women = 43.03.

The scatterplot suggests that the relation for men and women is about the same. However, the points for men appear to be slightly more scattered (in particular, the point at (51.9, 1867) gives the impression of greater scatter), so we might expect the correlation for men to be smaller than for women.

d) The value of the correlation is unchanged when the units of measurement are changed (see item 3 on page 132 of the textbook).

Exercise 3.26

a) Here is the scatterplot.

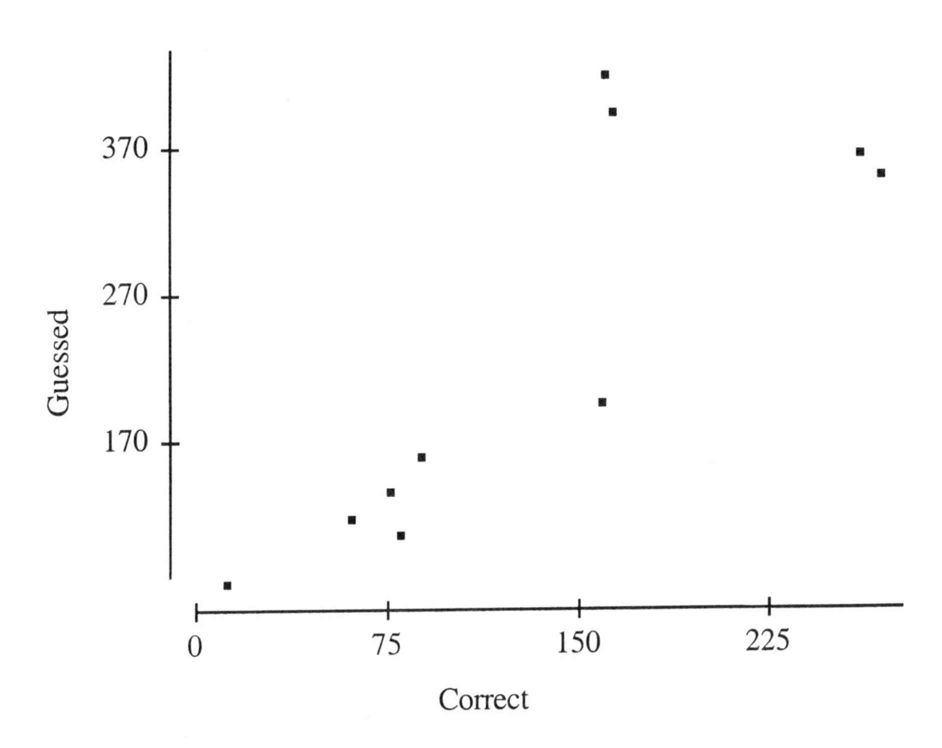

We find (displaying all the decimal places given by the calculator) that

$$r = \underline{0.8245015756}$$

The scatterplot shows a clear linear trend and positive association, and the value of r is consistent with this.

b) Changing the units of measurement by subtracting 100 from each guess would not change the correlation, but would make most of the guesses less than the correct value. Thus, the fact that all the guesses are higher than the correct value does not influence the correlation here. By the same reasoning, adding 100 to each of the guesses would not change r.

c) Here is the scatterplot with the points for spaghetti and snack cake circled.

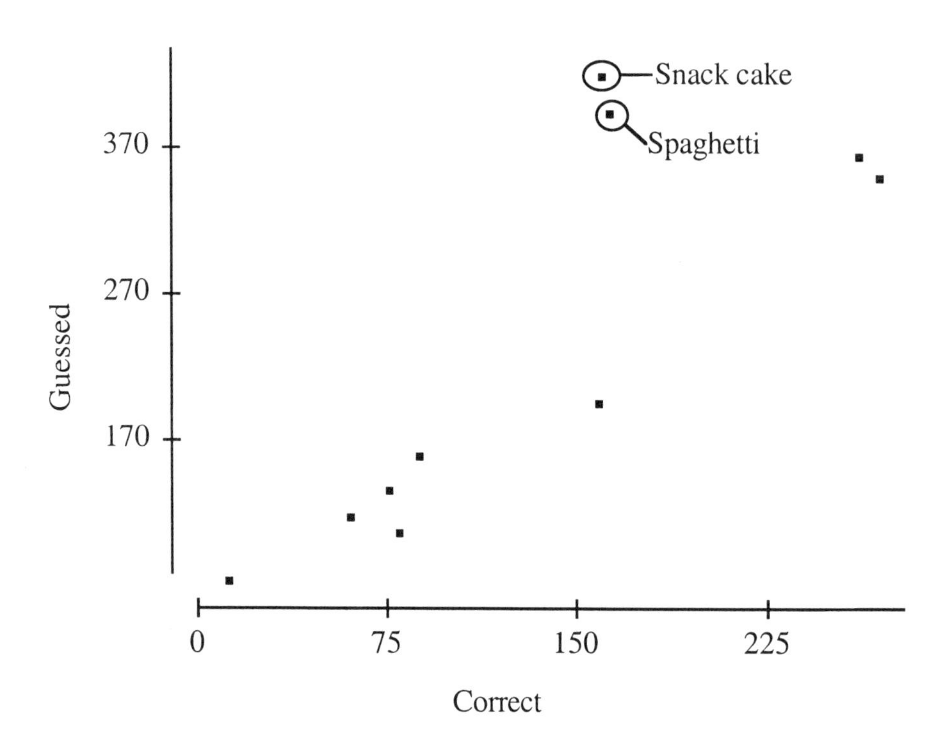

After deleting the values for spaghetti and snack cake, we find that

$$r = \underline{0.9837411881}$$

When the points corresponding to spaghetti and snack cake are removed from the scatterplot, the remaining points appear to lie much closer to a straight line. Therefore we would expect the correlation to increase when these values are deleted.

Exercise 3.28

If the correlation were close to zero, there would be no particular linear relationship. Good researchers would be just as likely as bad researchers to be good or bad teachers. The statement that "good researchers tend to be poor teachers, and vice versa" implies that the correlation is negative, not zero.

SECTION 3.3

OVERVIEW

If a scatterplot shows a linear relationship that is moderately strong as measured by the correlation, we can draw a line on the scatterplot to summarize the relationship. In the case where there is a response and an explanatory variable, the **least-squares regression** line often provides a good summary of this relationship. A straight line relating y to x has the form $y = a + bx$, where b is the **slope** of the line and a is the **intercept**. The least-squares regression line is the straight line $\hat{y} = a + bx$, which minimizes the sum of the squares of the vertical distances between the line and the observed values y. The formula for the slope of the least squares line is

$$b = r\frac{s_y}{s_x}$$

and for the intercept is $a = \bar{y} - b\bar{x}$, where $\bar{x}$ and $\bar{y}$ are the means of the x and y variables, s_x and s_y are their respective standard deviations, and r is the value of the correlation coefficient. Typically, the equation of the least-squares regression line is obtained by computer software or a calculator with a regression function.

Regression can be used to predict the value of y for any value of x. Just substitute the value of x into the equation of the least-squares regression line to get the predicted value for y.

Correlation and regression are clearly related, as can be seen from the equation for the slope, b. However, the more important connection is how the **coefficient of determination**, r^2, measures the strength of the regression. r^2 tells us the fraction of the variation in y that is explained by the regression of y on x. The closer r^2 is to 1, the better the regression describes the connection between x and y.

An examination of the **residuals** shows us how well our regression does in predictions. The difference between an observed value of y and the predicted value obtained by least-squares regression, $\hat{y}$, is called the residual.

$$\text{residual} = y - \hat{y}$$

Plotting the residuals is a good way to check the fit of our least-squares regression line. Features to look for in a **residual plot** are unusually large values of the residuals (outliers), nonlinear patterns, and uneven variation about the horizontal line through zero (corresponding to uneven variation about the regression line).

Also look for influential observations. **Influential observations** are individual points whose removal would cause a substantial change in the regression line. Influential observations are often outliers in the horizontal direction.

GUIDED SOLUTIONS

Exercise 3.33

KEY CONCEPTS: Drawing and interpreting the least-squares regression line

a) Perhaps the simplest way to draw a graph of the line given is to pick two convenient values for the variable weeks, substitute them into the equation of the least-squares regression line, and compute the corresponding value of pH predicted by the equation for each. This produces two sets of week and pH values. Each of these week and pH pairs corresponds to a point on the least-squares regression line. Simply plot these points on your graph and connect them with a straight line. Be sure that the horizontal axis of your graph corresponds to the explanatory variable or x value in the equation of the line (in this case, weeks) and the vertical axis to the response variable or y value in the equation (in this case, pH).

Convenient values for weeks might be 0 and 100. Complete the following.

For weeks = 0, pH = For weeks = 100, pH =

Next, plot these two sets of week and pH values as points on the axes that follow. Then connect them with a straight line. Alternatively, you can use your calculator or computer to produce a graph of the line.

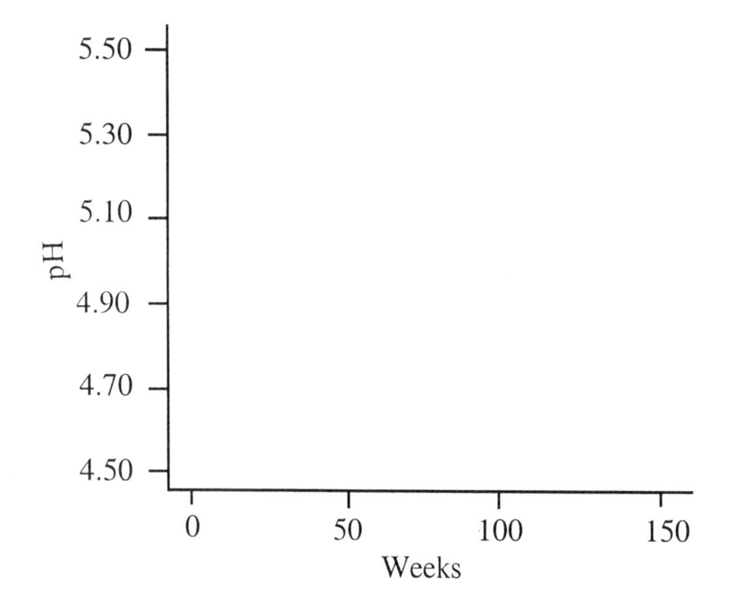

From the least-squares regression line, does pH increase or decrease as weeks increases? Is the association, therefore, positive or negative?

Write in plain language what this association means.

b) According to the equation of the least-squares regression line, namely,

$$\text{pH} = 5.43 - (0.0053 \times \text{weeks})$$

what is the value of pH at the beginning and end of the study?

For weeks = 1, pH =

For weeks = 150, pH =

c) For a line whose equation is

$$y = a + bx$$

the quantity b is the slope. Identify this quantity in the equation of the least-squares regression line.

Slope =

Explain clearly in writing what this slope says about the change in the pH of the precipitation in this wilderness area. (Writing out your explanation forces you to express your thoughts explicitly. If you cannot write a clear explanation, you might not adequately understand what the slope means.)

Exercise 3.37

KEY CONCEPTS: Scatterplots, correlation, the least-squares regression line, r^2

a) If you are using a calculator or statistical software, you should enter the data and use the calculator or software to create a scatterplot. It should look like the following.

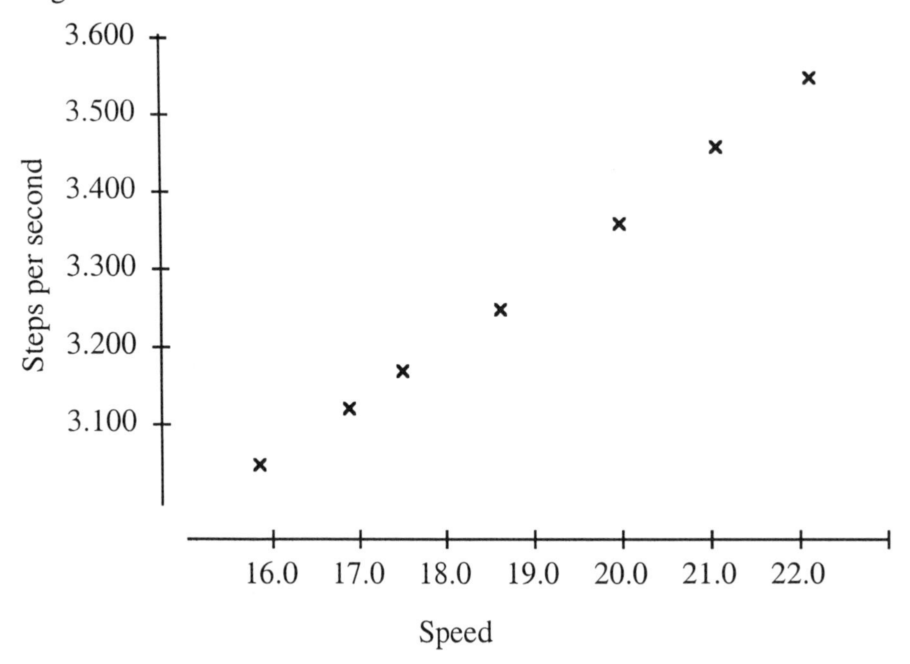

Although we are giving you many of the plots, it is a good idea to make sure you understand in each plot why one variable was designated the response and the other the explanatory variable.

b) What is the general trend in your scatterplot? Does it appear to be adequately described by a straight line or is curvature present?

Compute the correlation and enter it in the space provided.

$r =$ ________________________

c) Use a calculator (or statistical software) to compute the equation of the least-squares regression line. Write the equation in the space provided below.

Next, plot this line on your scatterplot in part a.

d) Compute r^2 and use it to explain whether running speed explains most of the variation in the number of steps a runner takes per second.

e) Is the least-squares regression line affected by which variable is the response and which is the explanatory? Is r^2 affected? You might want to refer to page 144 in your textbook.

Exercise 3.40

KEY CONCEPTS: Outliers and influential observations

a) If you have done exercise 3.12 (discussed previously in this guide), you have already sketched the scatterplot. If you haven't tried exercise 3.12 and you wish to sketch the plot by hand, use the axes below. Otherwise, use your calculator or computer to create the plot. Remember to circle the points for spaghetti and snack cake on your plot. Note that since we wish to predict guessed calories from true calories, true calories is the predictor variable while guessed calories is the response.

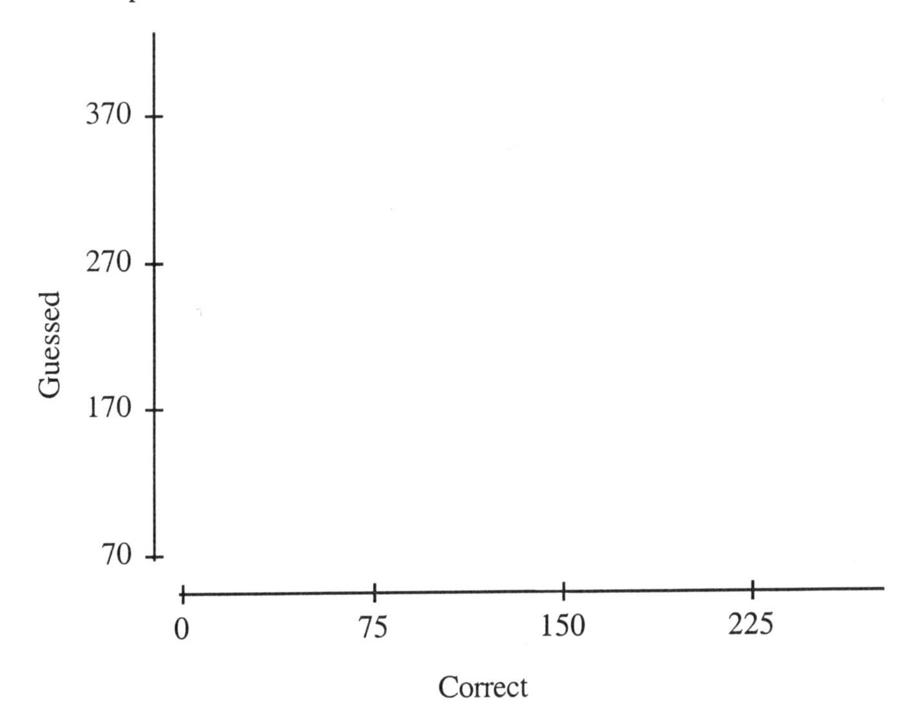

b) Use your calculator or statistical software to compute the equations of the least-squares regression lines. Record your results.

Equation of least-squares regression line for all ten points:

Equation of least-squares regression line after leaving out spaghetti and snack cake:

c) Sketch these two lines into your plot in part a. To help you distinguish them, make one of them dashed. To decide whether spaghetti and snack cake taken together are influential, ask yourself whether their removal resulted in a marked change in the equation of the least-squares regression line. You may wish to refer to Figure 3.17 in your textbook to give you a sense of how much change might be considered marked.

Exercise 3.46

KEY CONCEPTS: Scatterplots, r and r^2, the least-squares regression line from summary statistics, prediction, residuals, influential observations

Note: Because of the size of this data set, this problem is best done with a calculator or statistical software.

a) If you wish to make the scatterplot by hand, draw your scatterplot on the axes that follow. Otherwise, use your calculator or computer to create the plot. Be sure to label the axes correctly. Which variable is the predictor and which is the response?

b) Give the values of the correlation r and r^2.

$r =$

$r^2 =$

Next, describe the relationship between U.S. and overseas returns in words. Use r and r^2 to make your description more precise. You may find it helpful to refer to the facts about correlation in Section 2 of this chapter and, in particular, to Fact 4 about correlation.

c) You will want to use your calculator or statistical software to compute the equation of the least-squares regression line. Write the equation.

$\hat{y} =$

Next, draw this line in your plot in part a.

d) Use the equation of the least-squares regression line you found in part c to predict the return on overseas stock when the return on U.S. stocks is 23.0%.

Prediction =

How does your prediction compare to the 1996 overseas return of 6.4% when the U.S. return was 23.0%? Are you confident that predictions using the regression line will be quite accurate? Why? You may wish to think about what the answer to part b implies about prediction.

e) Circle the point in your plot in part a that has the largest residual. This point corresponds to what year (refer to the data in the statement of the problem to answer this)?

Year =

Are there any points in your plot that seem likely to be very influential? Which one(s)?

COMPLETE SOLUTIONS

Exercise 3.33

a) We find that

$$\text{for weeks} = 0,\ \text{pH} = 5.43 - (0.0053 \times 0) = 5.43$$

$$\text{for weeks} = 100,\ \text{pH} = 5.43 - (0.0053 \times 100) = 4.90$$

These two points are plotted below (using x as the plotting symbol), and the least-squares regression line is drawn to connect them.

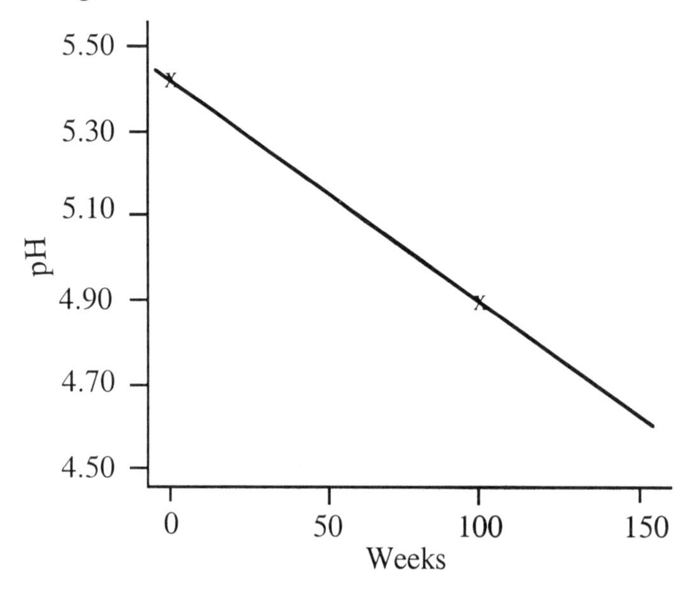

The line has a negative slope: As the variable weeks increases, pH decreases. This implies that the association is negative. In plain language, this association means that over the period of the study, as time passed, the pH of the precipitation tended to decrease and hence the acidity of the precipitation tended to increase.

b) From the equation of the least-squares regression line, namely,

$$\text{pH} = 5.43 - (0.0053 \times \text{weeks}),$$

we find that

$$\text{for weeks} = 1,\ \text{pH} = 5.43 - (0.0053 \times 1) = 5.4247$$

$$\text{for weeks} = 150,\ \text{pH} = 5.43 - (0.0053 \times 150) = 4.6350$$

c) From the equation of the least-squares regression line, namely

$$\text{pH} = 5.43 - (0.0053 \times \text{weeks})$$

we see that the slope is –0.0053. The slope tells us that, on average, for each week of the study period, the pH of the precipitation in this wilderness area decreased by 0.0053.

Exercise 3.37

a) A plot of the data is given in the guided solution.

b) A straight line "appears" to describe the data quite well and the association is positive. We compute the correlation to be (displaying all the decimal places given by the calculator)

$$r = \underline{0.9989877247}$$

Normally, we would round off this number to only two or three decimal places.

c) The least-squares regression line (as computed on a TI-83) is

$$\text{Steps per second} = 1.766077145 + 0.0802837878(\text{Speed}).$$

We have displayed all the decimal places given by our calculator, but normally we would round off these values to just a few decimal places.

A scatterplot with this line drawn in is given below.

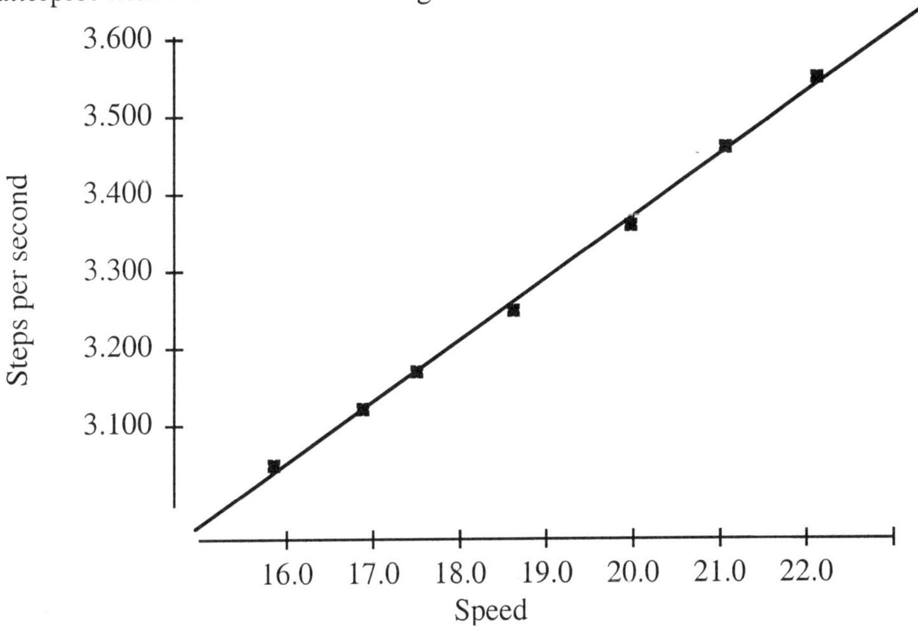

d) We find that $r^2 = 0.9979764741$. Recall that r^2 is the fraction of the variation in the values of the response (steps per second) that is explained by least-squares regression of the response (steps per second) on the explanatory variable (speed). Because r^2 is very close to 1, we see that most of the variation in the number of steps a runner takes per second is explained by running speed.

e) The least-squares regression line depends on which variable we call the response and which is the explanatory. See the example in Figure 3.11 (page 145) of your textbook. Thus, we cannot simply use the same line to predict running speed from a runner's steps per second. On the other hand, the correlation, and thus r^2, is not affected by which variable we call the response and which the explanatory.

Exercise 3.40

a) Here is the scatterplot with the points for spaghetti and snack cake circled.

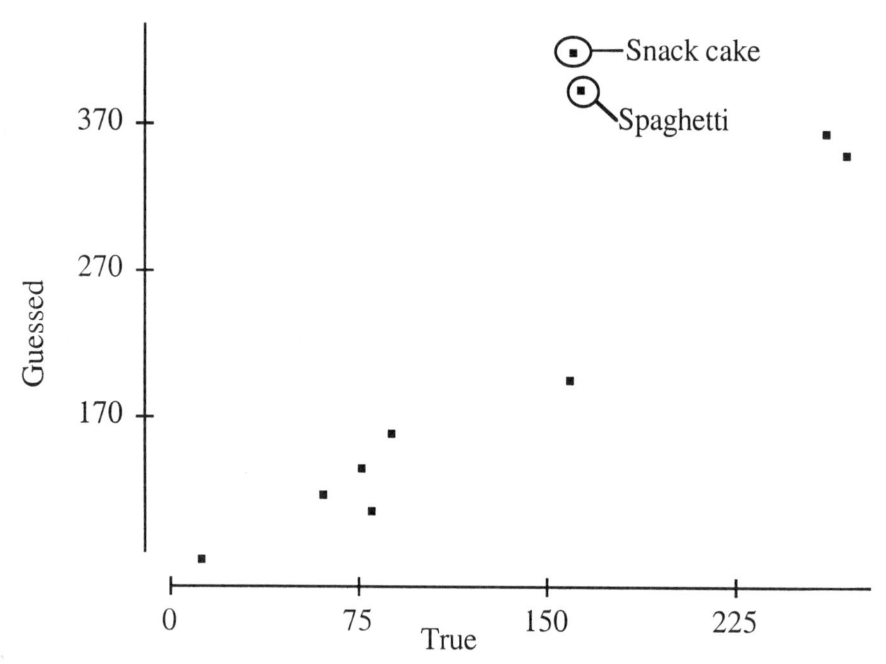

b) Here are the equations of the least-squares regression lines. You should obtain similar results using a calculator or statistical software, possibly after rounding.

For all ten data points:

$$\text{Guessed calories} = 58.588 + 1.304 \times (\text{true calories})$$

After leaving out spaghetti and snack cake:

$$\text{Guessed calories} = 43.881 + 1.147 \times (\text{true calories})$$

c) Here is our plot containing both lines.

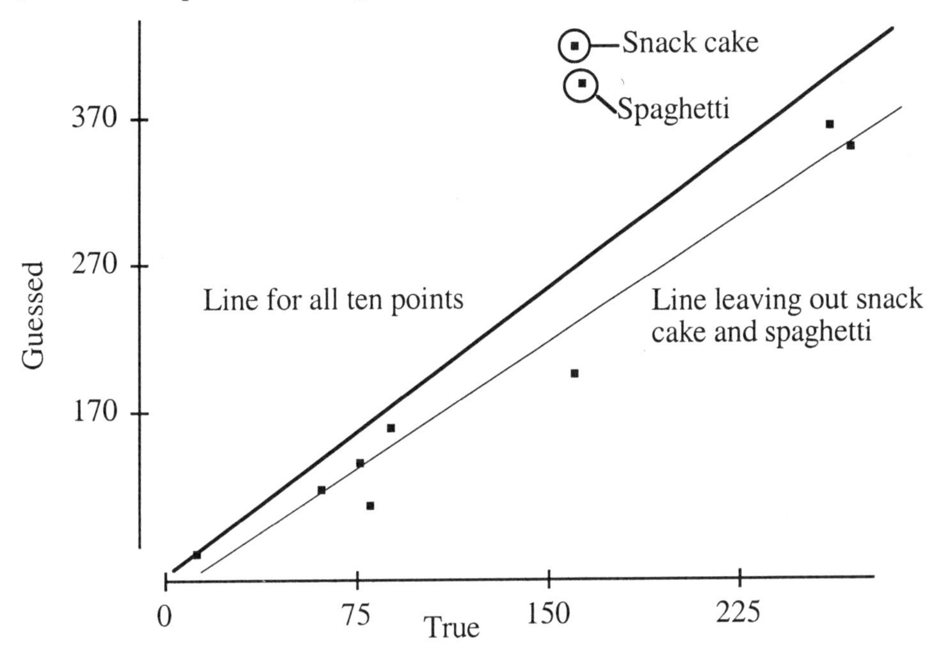

Removing snack cake and spaghetti causes the least-squares regression line to shift a noticeable amount (compare with Figure 3.17 in the textbook). We would consider spaghetti and snack cake taken together as influential.

Exercise 3.46

a) In this problem, we used statistical software to create our plots and carry out computations. Our scatterplot follows.

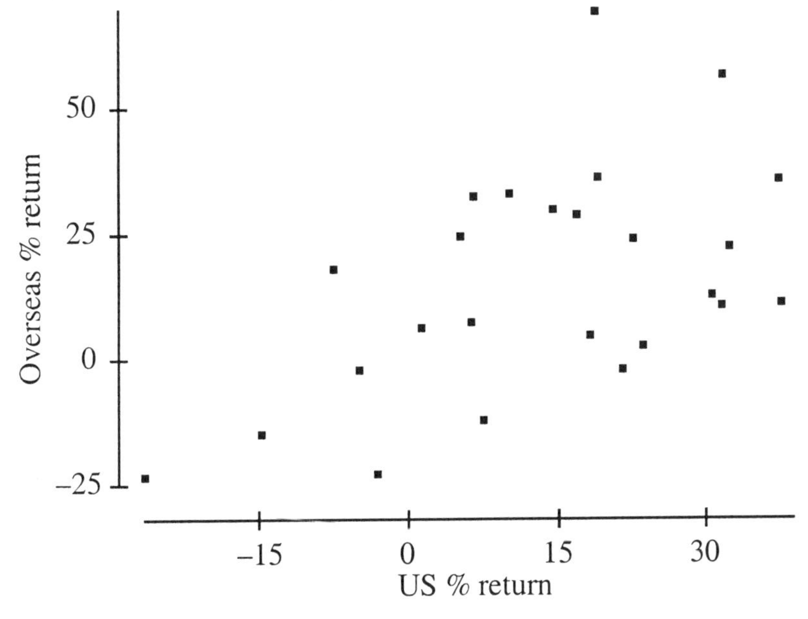

b) Our software computed the correlation r and r^2 to be

$$r = 0.522, \qquad r^2 = (0.522)^2 = 0.2725$$

Because the correlation is $r = 0.522$, there is a positive association between overseas and U.S. returns. In other words, above-average U.S. returns tend to be associated with above-average overseas returns. Likewise, below-average U.S. returns tend to be associated with below-average overseas returns. The strength of the association is not strong. Because $r^2 = 0.2725$, the fraction of the variation in overseas returns that is explained by the least-squares regression of overseas return on U.S. returns is only 27.25%.

c) We obtain the following equation of the least-squares regression line using our software (you should get a similar answer if you used a calculator):

$$\text{Overseas \% return} = 5.95456 + 0.713626 \times (\text{U.S. \% return})$$

We have displayed all the decimal places given by our statistical software, but it is probably better practice to round off to two or three decimal places.

The least-squares line is drawn in on the following scatterplot.

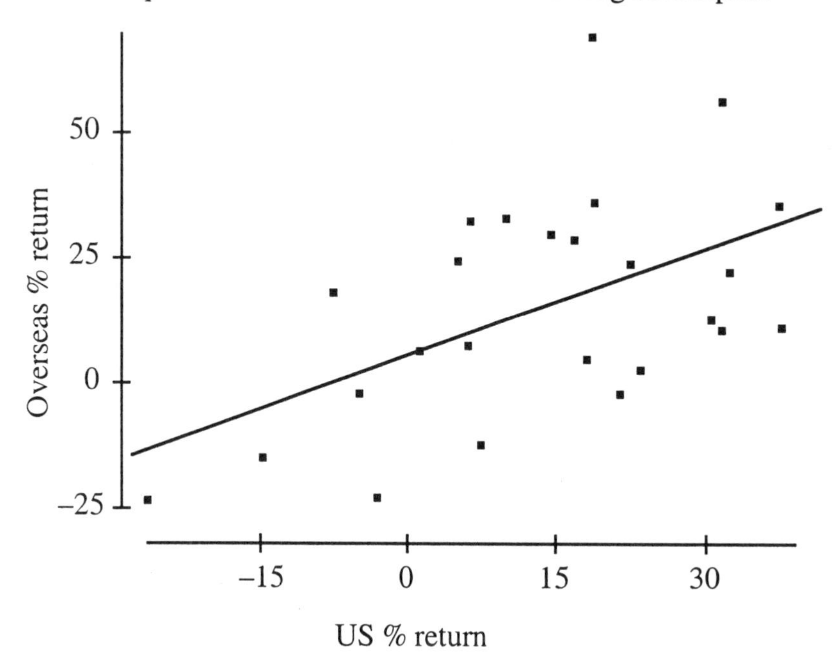

d) Using the equation of the least-squares regression line we obtained in part c, we predict

$$\text{Overseas \% return} = 5.95456 + 0.713626 \times (23.0) = 22.368\%$$

This does not compare well with the overseas return of 6.4% in 1996 when U.S. returns were 23.0%. We should not be overly surprised at the poor quality of our predictions. Remember, $r^2 = 0.2725$, so the fraction of the variation in overseas returns that is explained by the least-squares regression of overseas returns on U.S. returns is only 27.25%.

e) The point with the largest residual is circled in the plot below.

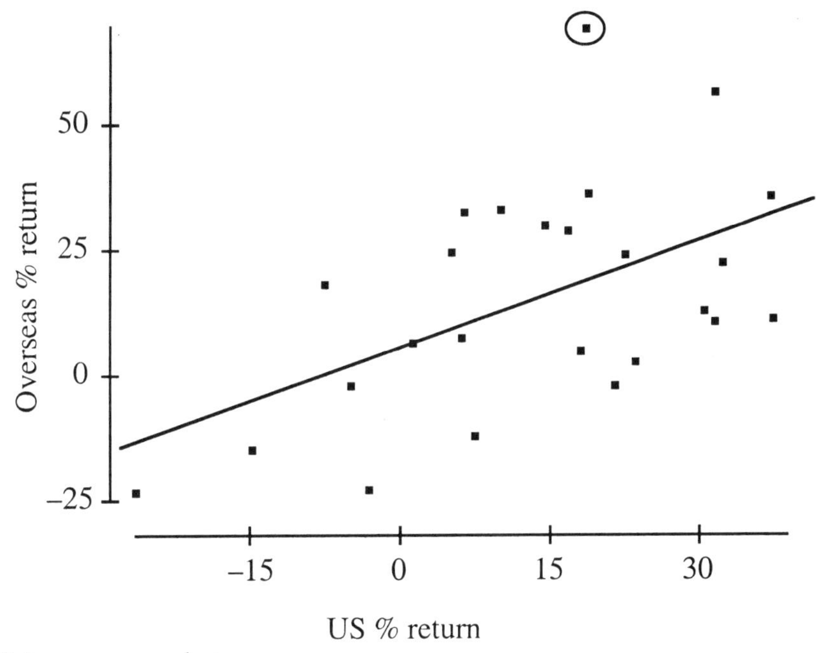

This corresponds to

Year = 1986.

The two point at the far left may be slightly influential.

CHAPTER 4

MORE ON TWO-VARIABLE DATA

SECTION 4.1

OVERVIEW

A variable displays **linear growth** when it increases by the addition of a fixed positive amount in each equal time period. A scatterplot of such a variable versus time will look approximately linear. The methods of least-squares regression can be used to explore linear growth of a variable over time. Plots of residuals can be used to detect deviations in the overall linear pattern.

A variable displays **exponential growth** when it increases by multiplication by a fixed amount greater than 1 in each equal time period. The form of the relationship between the response y and the explanatory variable x is $y = ab^x$. A scatterplot of the **logarithm** of y versus x will look approximately linear. Exponential growth over time can thus be explored by applying the methods of least-squares regression to the logarithm of the response. Plots of residuals can be used to detect deviations in the overall linear pattern of the logarithm, suggesting deviations in the overall pattern of exponential growth.

Nonlinear data may also be modeled by a **power function** $y = ax^b$ that passes through the origin. In this case, a scatterplot of the logarithm of y versus the logarithm of x will look approximately linear. The data can then be explored by applying the methods of least-squares regression to the logarithms of the response and explanatory variables. Plots of residuals can be used to detect deviations in the overall linear pattern of the logarithms, suggesting deviations in the overall pattern of the power function.

GUIDED SOLUTIONS

Exercise 4.1

KEY CONCEPTS: Exponential growth

a) If you wish to make the scatterplot by hand, draw your scatterplot on the axes that follow. Otherwise, use your calculator or computer to create the plot. In this case, there are only four points to plot, and drawing the plot by hand may be easier than creating it using a calculator or computer.

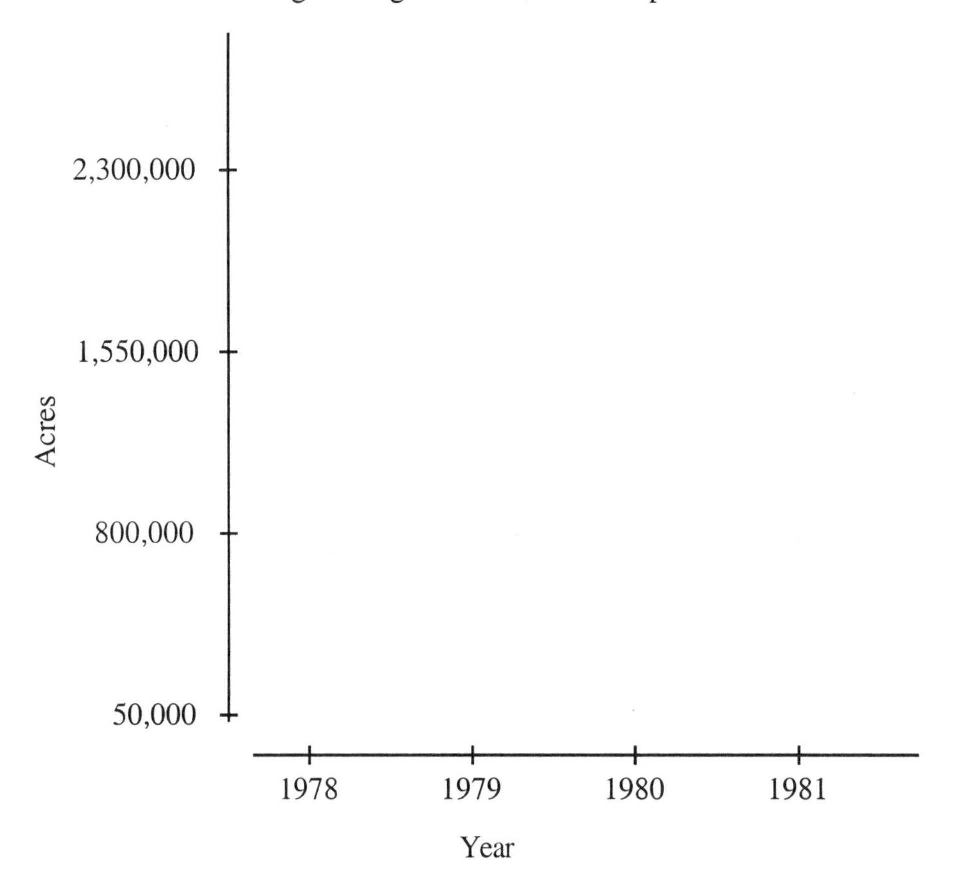

You should see a pattern that appears to be exponential.

b) Calculate the ratios and fill in the table that follows. We have completed the first entry to get you started.

Years	Ratio	
1979 to 1978	226260/63042	= 3.6
1980 to 1979	907075/226260	=
1981 to 1980	2826095/907075	=

c) First compute the logarithms (base 10) of Acres.

Acres	$\log_{10}$(Acres)
63042	
226260	
907075	
2826095	

Next, plot the logarithms against the year x using your calculator, computer, or the axes that follow.

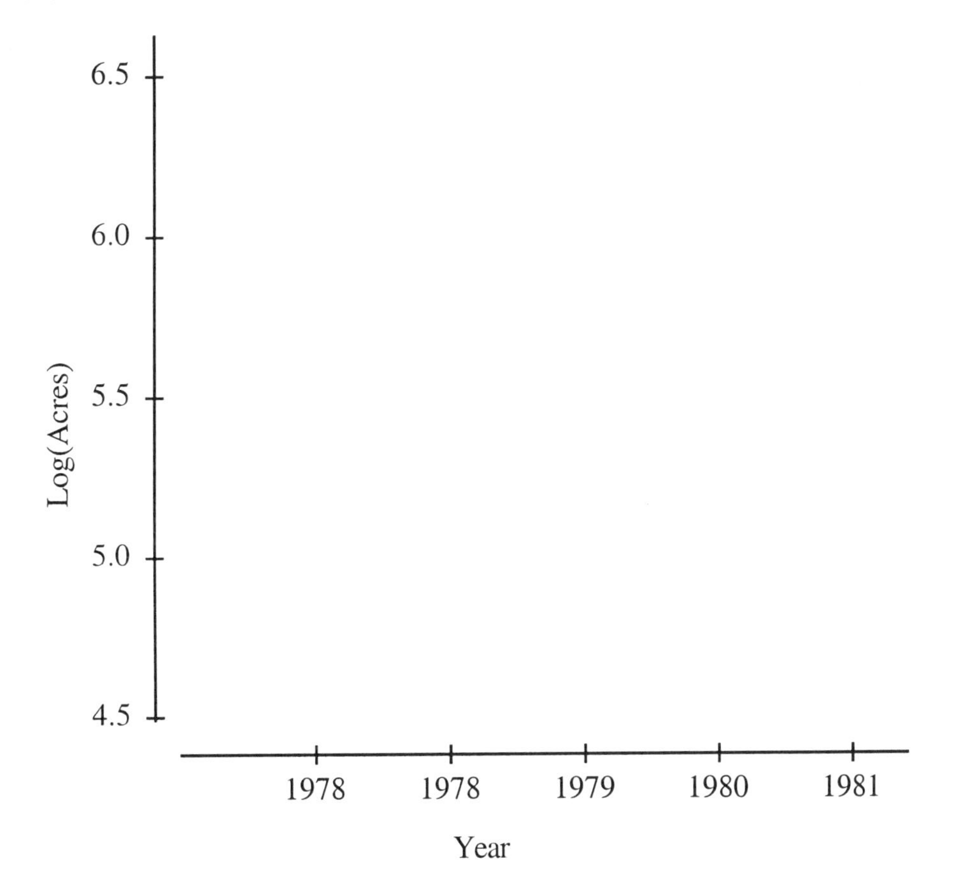

Is the pattern linear?

d) Use your calculator or computer to find the least-squares regression line of log(Acres) on year. Do your results match the equation

$$\log \hat{y} = -1094.51 + .5558 \times \text{year}$$

given in the statement of the problem?

e) Use your calculator or computer to construct a plot of the residuals against year. If you prefer to construct the plot by hand, use the axes below.

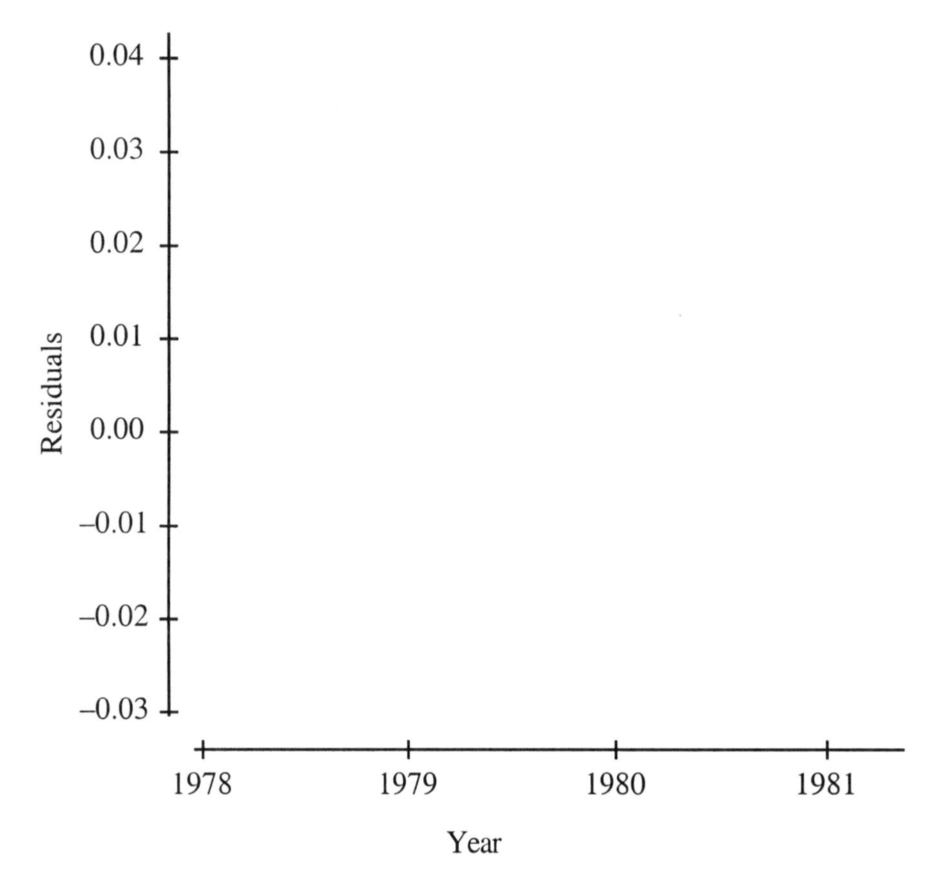

Next, interpret the plot.

f) Raise 10 to both sides of the equation to express $\hat{y}$ as an exponential equation. Write the result below.

$$\hat{y} =$$

Use your calculator, computer, or the plot below to graph the exponential curve.

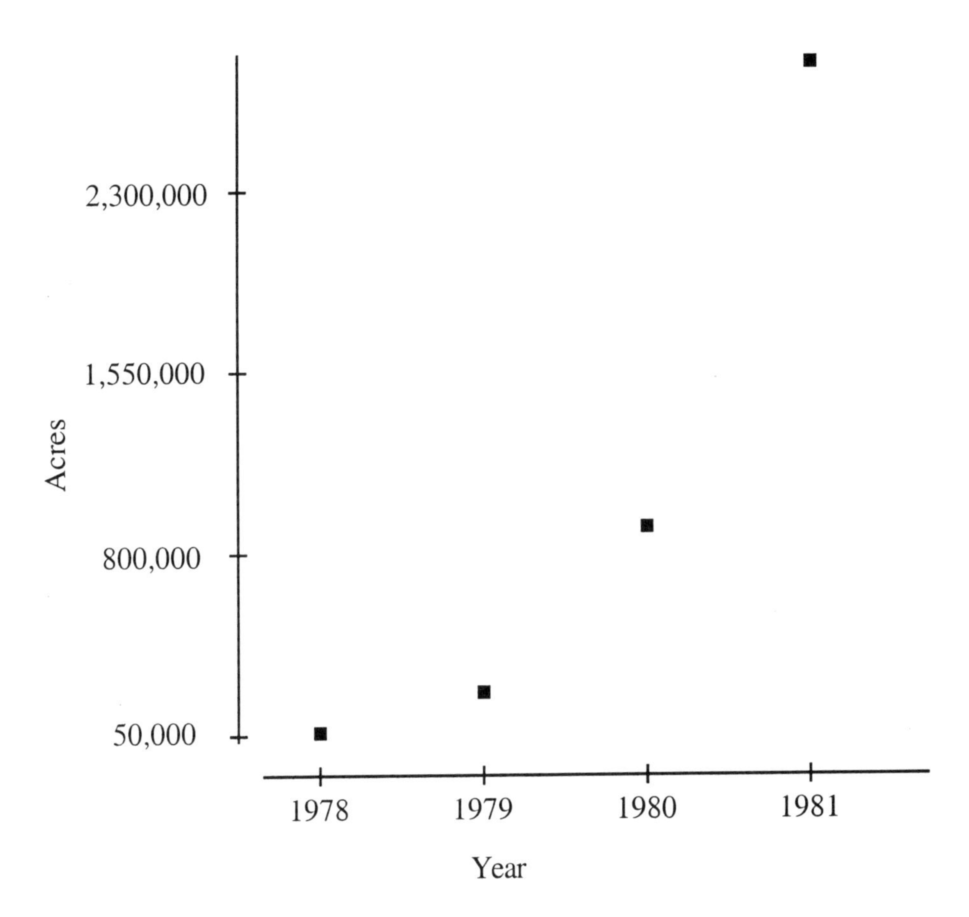

Describe the quality of the fit.

g) Using your model in part f, what would you predict the number of defoliated acres to be in 1982?

Predicted value of y for 1982 = ______________________

Exercise 4.17

KEY CONCEPTS: Power regression

a) Use your calculator, computer, or the axes below to make your plot.

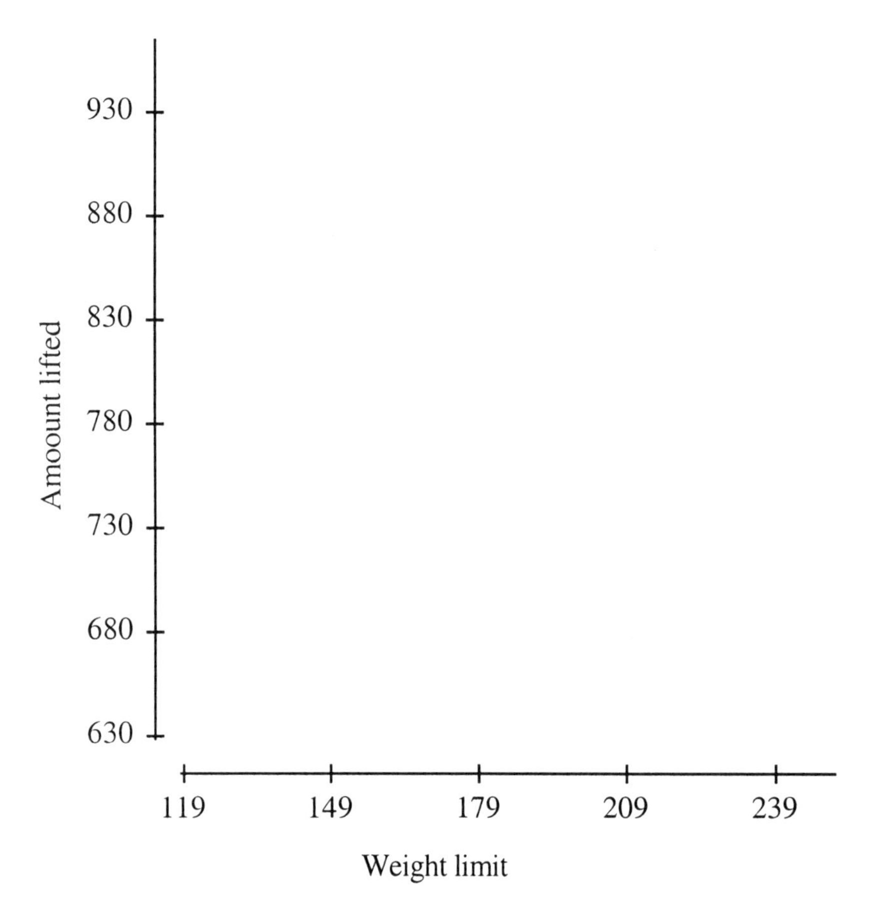

What sort of relationship do you notice?

b) Use QuadReg in the STAT/CALC menu. Remember to put the explanatory variable (Weight limit) in L_1 and the response (Amount lifted) in L_2. Write the equation in the space provided.

$\hat{y}$ =

To assess the fit, plot your quadratic regression curve on the scatterplot you produced in part a and construct a residual plot. For the residual plot, use the axes below if you are producing the plot by hand. What do you observe?

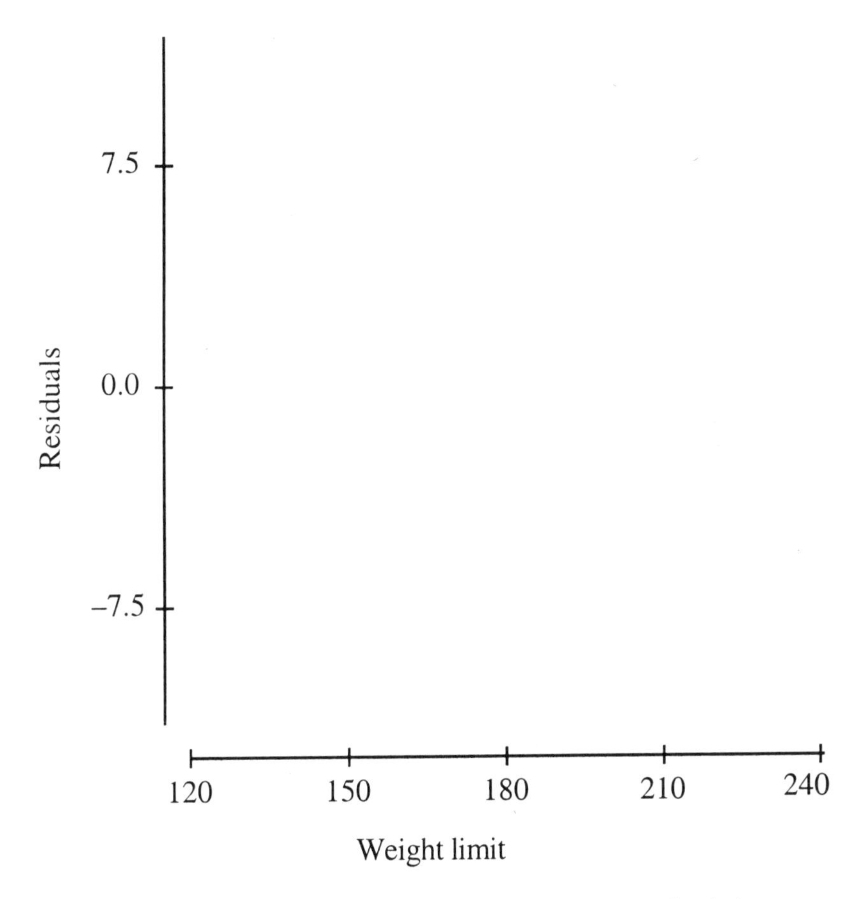

c) To help you answer this, compute the amount lifted that your model would predict for a weight limit of 0 pounds and a weight limit of 300 pounds

predicted amount lifted for a weight limit of 0 pounds =

predicted amount lifted for a weight limit of 300 pounds =

Comment.

d) What aspects of least-squares regression are affected by a change of units to both the response and explanatory variable? Consider the constant term, the coefficient of the linear term, and the coefficient of the quadratic term separately. How would the change affect your plots?

Complete Solutions

Exercise 4.1

a)

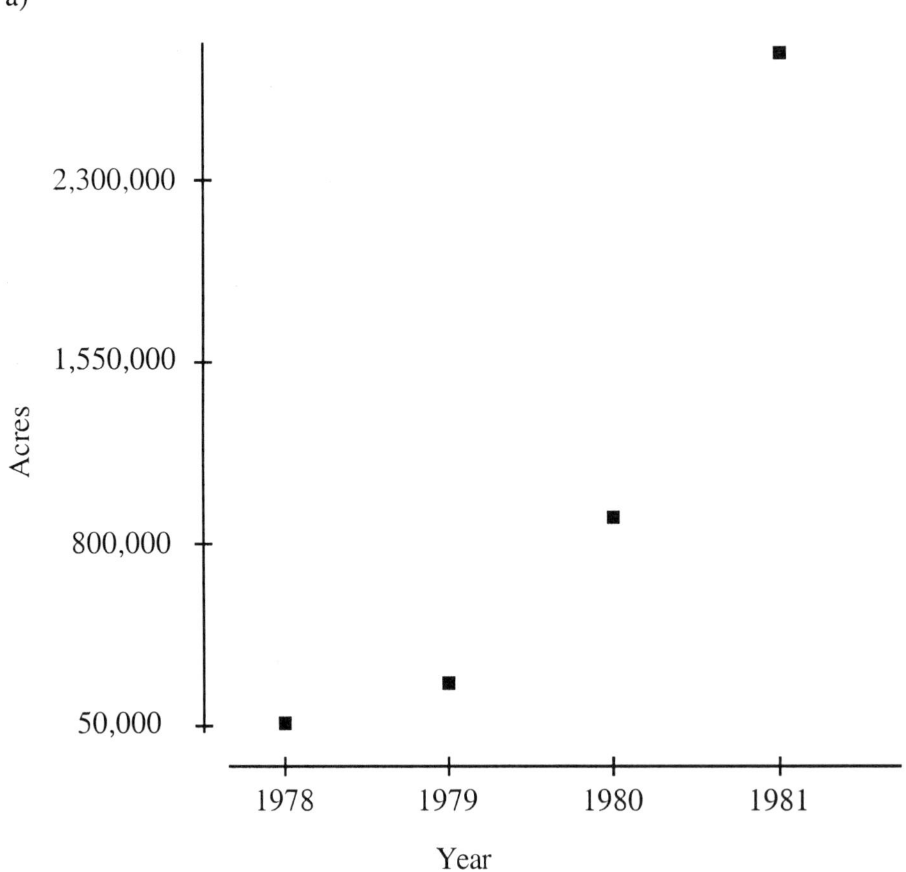

The pattern of growth appears to be exponential.

b)

Years	Ratio	
1979 to 1978	226260/63042	= 3.6
1980 to 1979	907075/226260	= 4.0
1981 to 1980	2826095/907075	= 3.1

The ratios are between 3 and 4.

c) The logarithms are (to two decimal places)

Acres	$\log_{10}$(Acres)
63042	4.80
226260	5.35
907075	5.96
2826095	6.45

and a plot of these against year is

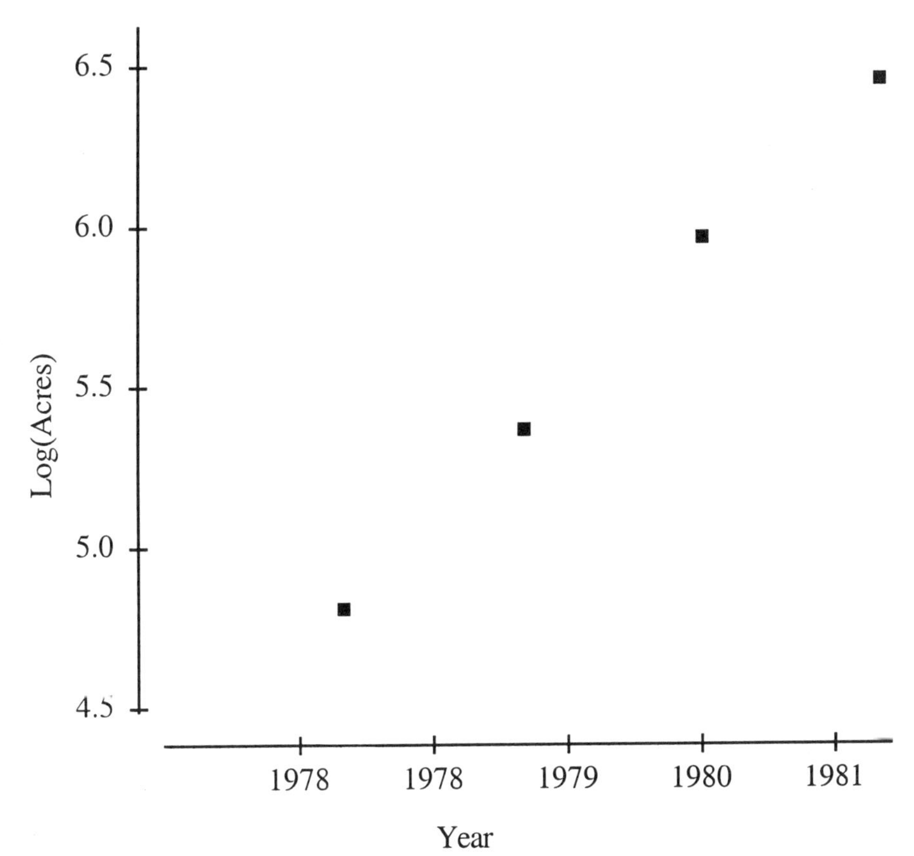

The pattern is very close to linear.

d) Using a calculator, we obtain the least-squares line

$$\log \hat{y} = -1094.50708 + .55577 \times \text{year},$$

which agrees with the equation given in the statement of the problem.

e) Here is the residual plot.

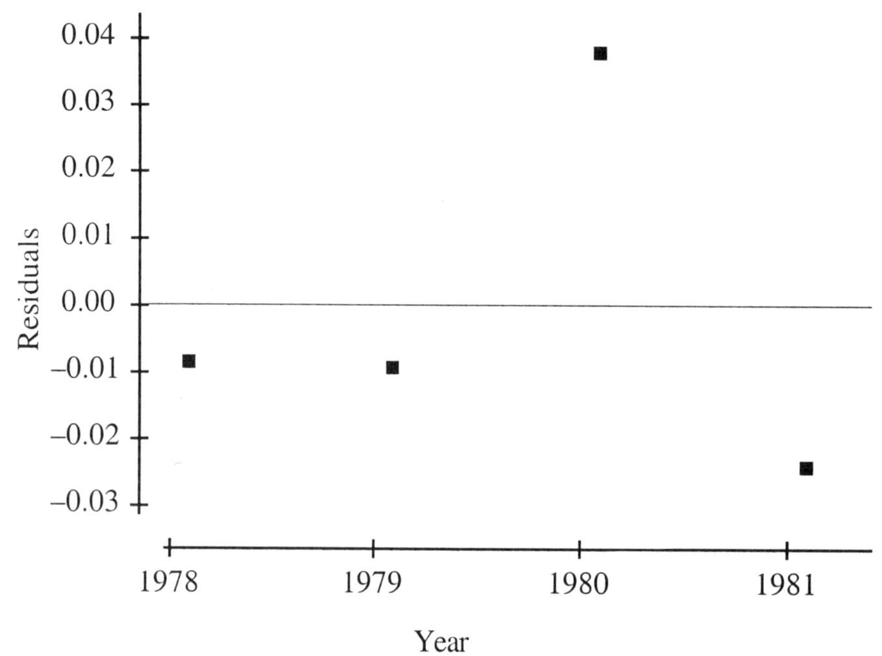

There are no striking patterns that would suggest that the line is not a reasonable fit. With so few points, only very striking patterns would indicate a problem.

f) The inverse transformation gives $\hat{y} = 10^{-1094.51+.55577 \times \text{year}}$. Below is the plot of this equation is superimposed on the scatterplot of the original data.

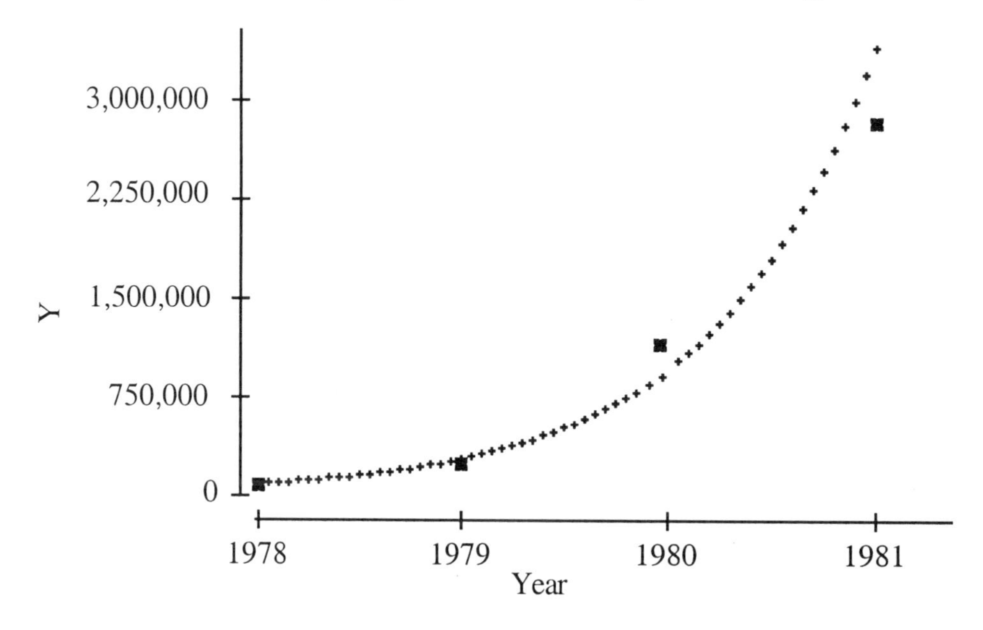

The fit appears to be satisfactory, but notice the discrepancy between the equation and the actual values in 1980 and 1981. The fit is not as good in later years.

g) We would predict that

$$\hat{y} = \text{Predicted value of } y \text{ for } 1982 = \underline{10{,}620{,}378.63}$$

Exercise 4.17

a) The scatterplot is given below. There appears to be a distinct curved relationship that is flattening out as weight limit increases.

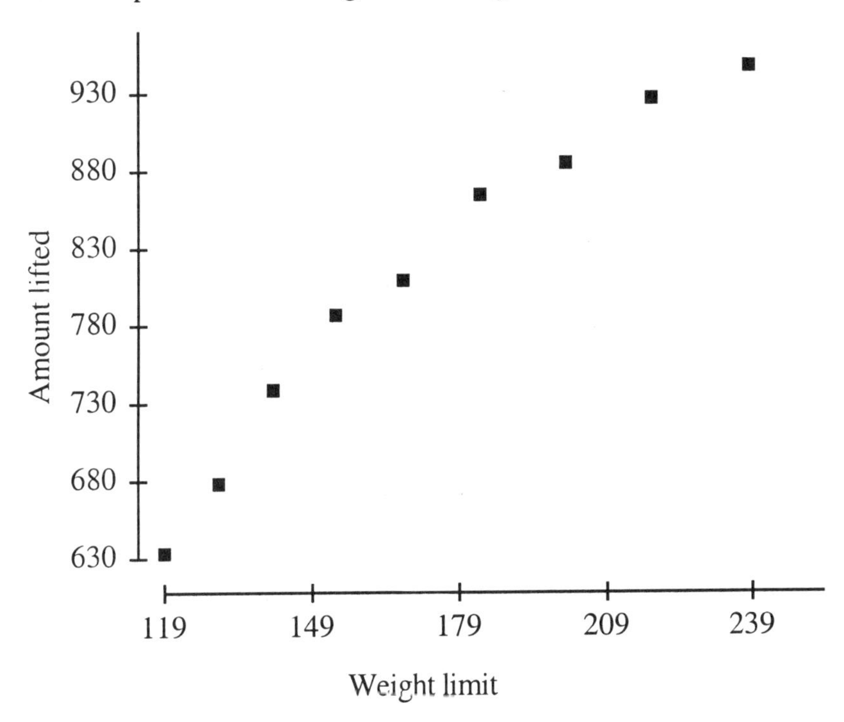

b) Using a TI-83, we obtain the following quadratic model

$$\hat{y} = -132.9841 + 8.4121 \times (\text{Weight limit}) - 0.0163 \times (\text{Weight limit})^2$$

A plot of this curve superimposed on the scatterplot of the data follows.

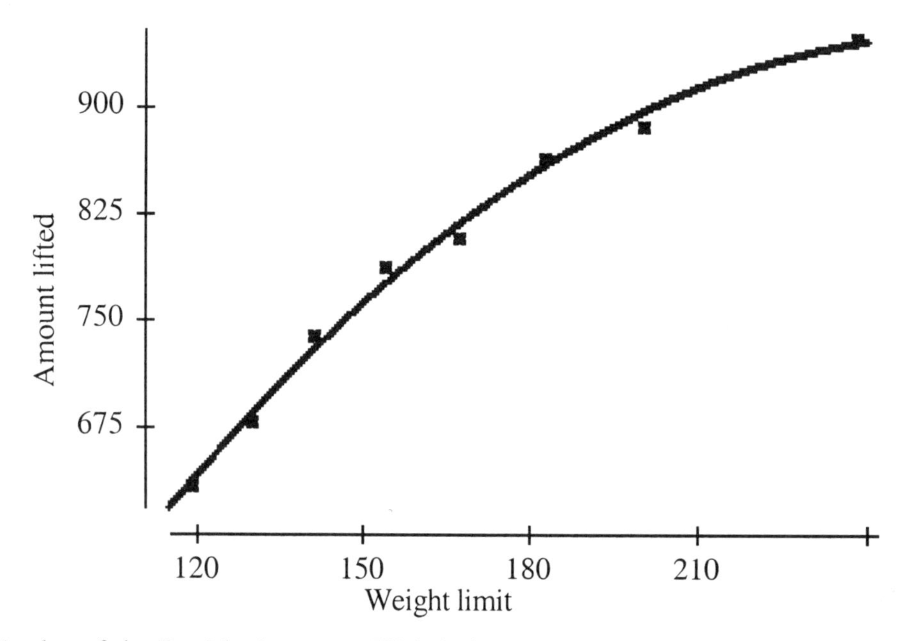

A plot of the Residuals versus Weight limit is

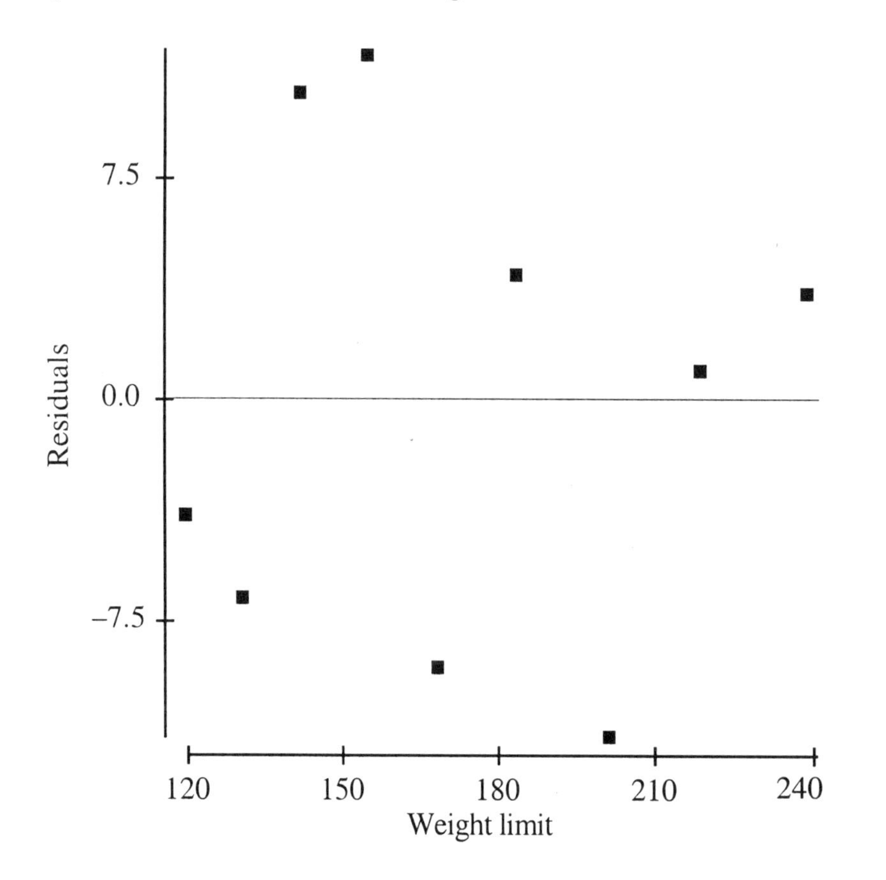

The data appear to fit the model reasonably well. There are no unusual patterns in the residual plot.

c) Extrapolation outside the range of data is always dangerous. To see this we compute

predicted amount lifted for a weight limit of 0 pounds =

$$-132.9841 + 8.4121 \times (0) - 0.0163 \times (0)^2 = -132.98 \text{ pounds}$$

predicted amount lifted for a weight limit of 300 pounds =

$$-132.9841 + 8.4121 \times (300) - 0.0163 \times (300)^2 = 923.6 \text{ pounds.}$$

The prediction at 0 pounds of a negative weight lifted makes no sense. The predicted weight lifted at 300 pounds is less that lifted at 238 pounds. One would expect that the weight lifted at 300 pounds to be at least as much as that at 238 pounds because lifters weighing up to 300 pounds are larger—and hence potentially stronger—than those in smaller weight classes. This underscores the unreliability of using our model outside the range of the data.

d) The coefficients of the constant and quadratic terms in the quadratic regression equation would change. While the scales of the axes in our plots also would change, the shapes of the plots would remain the same.

SECTION 4.2

OVERVIEW

Regression and correlation are potentially powerful ways of describing the relation between two variables. Here are some limitations.

Watch out for the following:

- Do not **extrapolate** beyond the range of the data.
- Be aware of possible **lurking variables**.
- Are the data averages or from individuals? Using **averaged data** usually leads to overestimating the correlations.
- Most of all, remember that *association is not causation!* Just because two variables are correlated doesn't mean one causes changes in the other.

GUIDED SOLUTIONS

Exercise 4.19

KEY CONCEPTS: Scatterplots, least-squares regression, r^2, extrapolation

a) Using your calculator, computer, or the axes provided, draw the scatterplot. Which is the response variable and which the explanatory variable? Which goes on the vertical axis and which goes on the horizontal axis? Be sure to label axes clearly if you draw your plot below.

Use your calculator or statistical software to find the least-squares regression line. Write the equation.

b) What quantity in the least-squares regression line represents the average decline per year in farm population during this period? What is its value? In answering this question, remember to keep in mind the units of the variable Population.

What quantity represents the percent of the observed variation in farm population that is accounted for by linear change over time? What is the value of this quantity?

c) Use the equation of the least-squares regression line you computed in part a to answer this question. Again, remember the units of the variable Population in reporting your answer. Is your prediction reasonable? Why?

Exercise 4.21

KEY CONCEPTS: Correlations based on averaged data

Regarding whether the correlation would increase or decrease if we had data on the individual stride rates of all 21 runners, the following quote from the textbook is relevant: "A correlation based on averages over many individuals is usually higher than the correlation between the same variables based on data for individuals."

Exercise 4.27

KEY CONCEPTS: Lurking variables

Ask yourself, what sorts of students are likely to study foreign language for at least two years. How are such students likely to do in other subjects?

COMPLETE SOLUTIONS

Exercise 4.19

a) Here is a scatterplot of the data with year the explanatory variable and population the response.

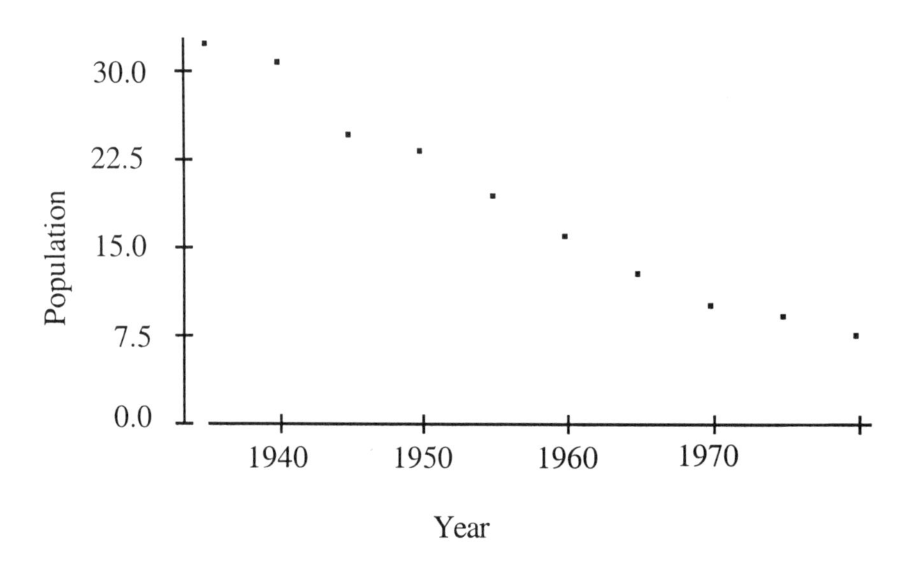

We find that the least-squares regression line is

$$\text{Population} = 1166.93 - 0.58679 \times \text{Year}$$

b) The slope of the least-squares regression line indicates the decline in farm population per year over the period represented by the data. This is a decline of 0.58679 million people per year or 586,790 people per year. The percent of the observed variation in farm population accounted for by linear change over time is determined by the value of r^2, which is 0.977 (calculated using software). The desired percent is therefore 97.7%.

c) In 1990, the regression equation predicts the number of people living on farms to be about

$$\hat{y} = 1166.93 - (0.58679 \times 1990) = -0.782 \text{ million.}$$

This result is unreasonable because population cannot be negative. This is an example of the dangers of extrapolation!

Exercise 4.21

If we had the data on the individual stride rates of all 21 runners, we would expect the correlation to decrease.

Exercise 4.27

The explanatory variable in this study is whether or not a student has studied a foreign language for at least two years. The response variable is scores on an English achievement test.

Probably the most important lurking variable is "quality of student" (how hardworking the student is, innate intelligence, innate language ability). Students who are willing to take at least two years of foreign language are likely to be more serious or talented students who are likely to work hard or do well in other subjects.

SECTION 4.3

OVERVIEW

This section discusses techniques for describing the relationship between two or more categorical variables. To analyze categorical variables, we use counts (frequencies) or percents (relative frequencies) of individuals who fall into various categories. **A two-way table** of such counts is used to organize data about two categorical variables. Values of the **row variable** label the rows that run across the table, and values of the **column variable** label the columns that run down the table. In each cell (intersection of a row and column) of the table, we enter the number of cases for which the row and column variables have the values (categories) corresponding to that cell.

The **row totals** and **column totals** in a **two-way table** give the marginal distributions of the two variables separately. It is usually clearest to present these distributions as percents of the table total. **Marginal distributions** do not give any information about the relationship between the variables. **Bar graphs** are a useful way of presenting these marginal distributions.

The **conditional distributions** in a two-way table help us to see relationships between two categorical variables. To find the conditional distribution of the row variable for a specific value of the column variable, look only at that one column in the table. Express each entry in the column as a percent of the column total. There is also a conditional distribution of the row variable for each column in the table. Comparing these conditional distributions is one way to describe the association between the row and column variables, particularly if the column variable is the explanatory variable. When the row variable is explanatory, find the conditional distribution of the column variable for each row and compare these distributions. Side-by-side bar graphs of the conditional distributions of the row or column variable can be used to compare these distributions and describe any association that may be present.

Data on three categorical variables can be presented as separate two-way tables for each value of the third variable. An association between two variables that holds for each level of this third variable can be changed, even reversed, when the data are combined by summing over all values of the third variable. **Simpson's paradox** refers to such reversals of an association.

GUIDED SOLUTIONS

Exercise 4.30

KEY CONCEPTS: Marginal distribution

Simply use your calculator to sum the four entires in the 35 to 54 column of Table 4.6. What do you find?

Exercise 4.31

KEY CONCEPTS: Marginal distribution

The marginal distribution of age can be found from the totals in each age group given in the bottom row of Table 4.6. Each value in this row must be divided by the total number of persons represented by the table found in the lower right corner. Next, do the actual calculations, completing the following table.

	Age Group		
	25–34	35–54	55 and older
Fractions			
Percent			

Exercise 4.33

KEY CONCEPTS: Describing relationships

Use the counts in the "Did not complete high school" row to find the requested percents. To organize your calculations, you may want to fill in the following table.

	Age Group		
	25–34	35–54	55 and older
Fractions			
Percent			

Next, based on the percents you calculated, draw the bar graph.

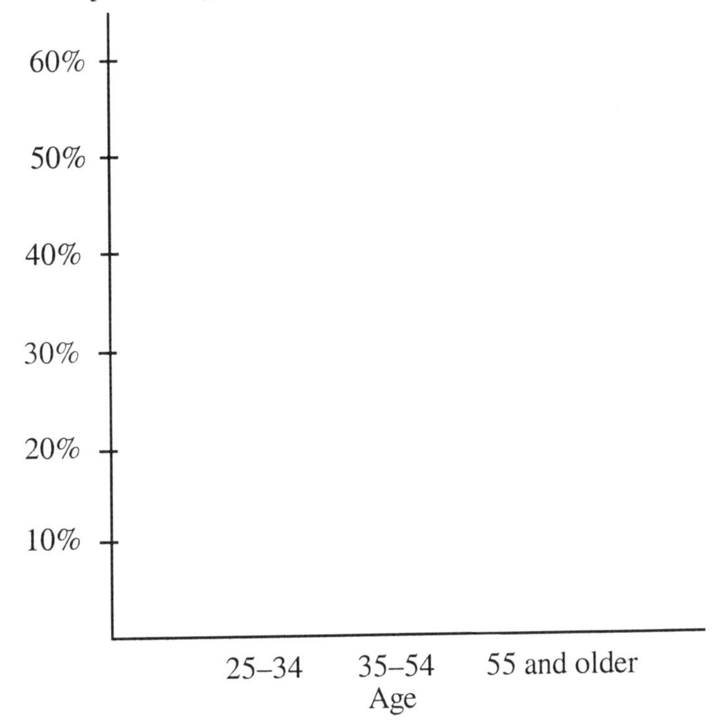

Describe what the data show.

Exercise 4.40

KEY CONCEPTS: Two-way tables, Simpson's paradox

a) Add corresponding entries in the two tables and enter the sums in the table.

	Admit	Deny
Male		
Female		

b) Convert your table in part a to one involving percentages of the row totals.

	Admit	Deny
Male		
Female		

c) Next, repeat the type of calculations you did in part b for each of the original tables.

Business

	Admit	Deny
Male		
Female		

Law

	Admit	Deny
Male		
Female		

d) To explain the apparent contradiction observed in part c, consider which professional school is easier to get into and which professional school males and females tend to apply to. Write your answer in plain English in the space provided. Avoid jargon and be clear!

Exercise 4.63

KEY CONCEPTS: Conditional distributions in two-way tables

Begin by converting the counts in the table to percentages of the column totals (in other words, compute the conditional distributions given gender).

	Male	Female
Firearms		
Poison		
Hanging		
Other		

Next, write a brief account of differences in suicide between men and women.

Exercise 4.65

KEY CONCEPTS: Marginal and conditional distributions, describing relationships

a) Use your calculator to compute this sum. What do you observe?

b) To compute the marginal distribution, compute the column totals in the following table.

	Single	Married	Widowed	Divorced
Total				

Fill in the following table by dividing the column totals by the grand total of 95,833.

	Single	Married	Widowed	Divorced
Fraction				
Percent				

Next, draw the bar graph.

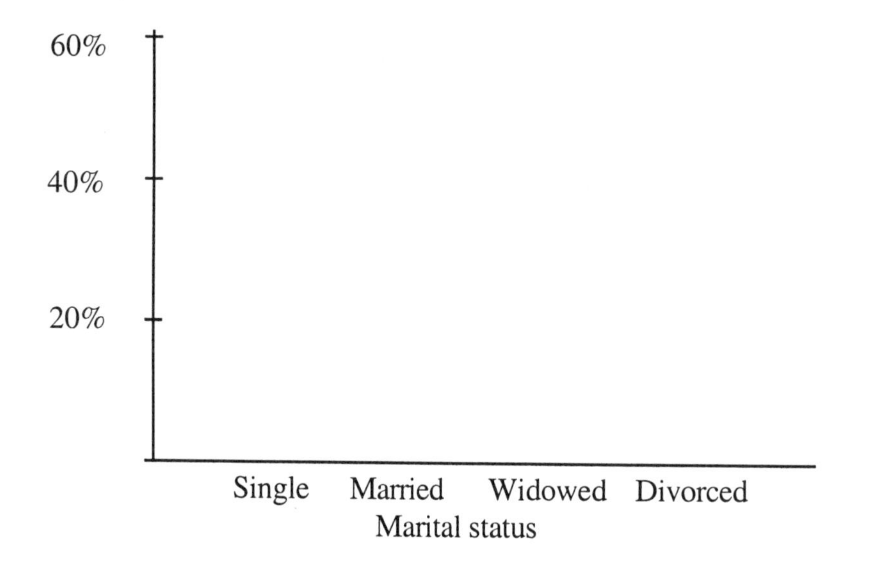

c) The conditional distributions are obtained by dividing the entry in each row by the total for the row and converting this fraction to a percent. Enter your results in the table.

	Single	Married	Widowed	Divorced
18–24				
40–64				

Use the space below for your description.

d) Fill in the percents corresponding to the conditional distribution of ages among single women.

	18–24	25–39	40–64	≥65
Percents				

What age groups should your magazine aim to attract? (Assume that "single" means "never married.")

COMPLETE SOLUTIONS

Exercise 4.30

The sum of the entries in the 35–54 column is 73,026. This is less than the value displayed in the table. Presumably the difference is due to rounding of the individual table entries.

Exercise 4.31

The marginal distribution of age can be found from the totals in each age group given in the bottom row of Table 4.6. Each value in this row must be divided by the total number of persons represented by the table found in the lower right corner, which is 166,438 thousand. The result is

	Age Group		
	25–34	35–54	55 and older
Fractions	$\frac{41,388}{166,438}$	$\frac{73,028}{166,438}$	$\frac{52,022}{166,438}$
Percent	24.86	43.88	31.26

Exercise 4.33

The percent of people in each age group who did not complete high school is determined by dividing the counts in the "Did not complete high school" row of the table by the total for each age group column. The result is

	Age Group		
	25–34	35–54	55 and older
Fractions	$\frac{5,325}{41,388}$	$\frac{9,152}{73,028}$	$\frac{16,035}{52,022}$
Percent	12.87	12.53	30.82

Following is a bar graph displaying these percents.

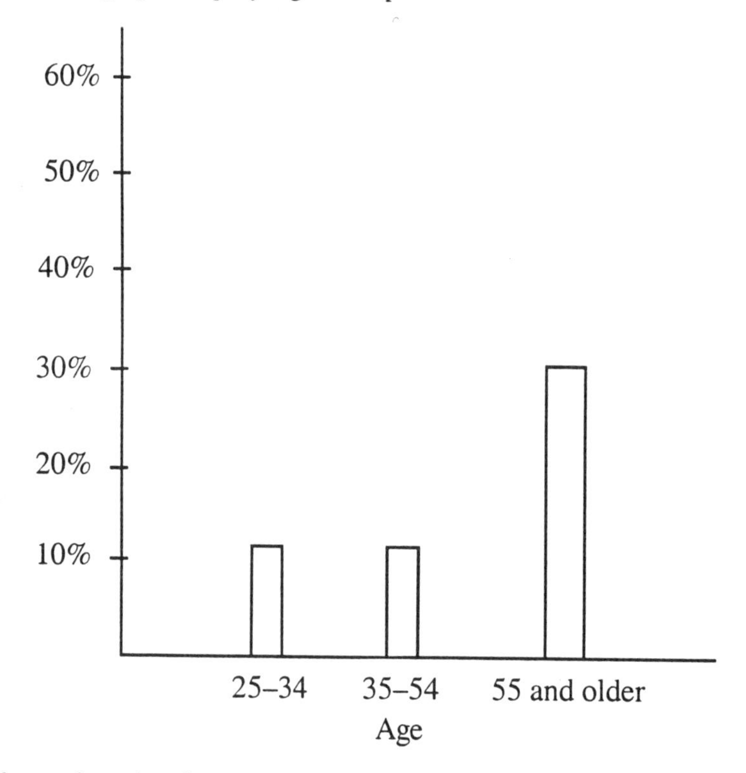

The data show that the the percentage who did not complete high school is essentially the same for 25–34 and 35–54 year olds, but is much higher for those 55 and older.

Although not requested, note that one possible explanation is that a greater emphasis on education (completion of high school) existed beginning in the 1950s (with concerns about the cold war and the space race). This is reflected in the higher percentage of people 55 and older who did not complete high school.

Exercise 4.40

a) Here is the desired two-way table.

	Admit	Deny
Male	490	210
Female	280	220

b) First, we add a column containing the row totals to the table in part a.

	Admit	Deny	Total
Male	490	210	700
Female	280	220	500

Next, we convert the table entries to percents of the row totals. We divide the entries in the first row by 700 and express the results as percents. We divide the entries in the second row by 500 and express these as percents.

	Admit	Deny
Male	70%	30%
Female	56%	44%

We see that Wabash admits a higher percent of male applicants.

c) We repeat the calculations in part b, but for each of the original two tables.

Business

	Admit	Deny	Total
Male	480	120	600
Female	180	20	200

Law

	Admit	Deny	Total
Male	10	90	100
Female	100	200	300

Converting entries to percents of the row totals yields

Business

	Admit	Deny
Male	80%	20%
Female	90%	10%

	Admit	Deny
Male	10%	90%
Female	33.3%	66.7%

We see that each professional school admits a higher percentage of female applicants.

d) Although both professional schools admit a higher percentage of female applicants, the admission rates are quite different. Business admits a high percentage of all applicants; it is easier to get into the business school. Law admits a lower percentage of applicants; it is harder to get into the law school Most of the male applicants to Wabash apply to the business school with its easy admission standards. Thus, overall, a high percentage of males are admitted to Wabash. The majority of female applicants apply to the law school. Because it has tougher admission standards, this makes the overall admission rate of females appear low, even though more females are admitted to both schools!

Exercise 4.63

We convert the table to percents of males and females by dividing by the column totals (the conditional distributions given gender).

	Male	Female
Firearms	65.87%	42.06%
Poison	13.03%	35.64%
Hanging	14.91%	12.23%
Other	6.19%	10.08%

A few things are striking. About four times as many men (24,724) as women (6,182) commit suicide. Firearms are the most popular method for both sexes, but men use firearms more frequently than women. Poison is nearly as popular as firearms for women. Poison is not as popular for men as it is for women. This may suggest that women have a greater preference for slower, more passive methods (such as poison) than men. Men seem to prefer quicker methods (firearms).

Exercise 4.65

a) The sum of the entries in the 18–24 row is actually

$$9{,}008+3{,}352+8+257 = 12{,}625$$

This differs from the entry in the Total column. This discrepancy is undoubtedly due to rounding off individual table entries to the nearest thousand women.

b) The distribution of marital status for all adult women is obtained by dividing the individual column totals by the table total given in the lower right corner of the table, then converting these fractions to percents.

	Single	Married	Widowed	Divorced
Total	18,541	56,838	11,290	9,161

	Single	Married	Widowed	Divorced
Fraction	$\frac{18{,}541}{95{,}833}$	$\frac{56{,}838}{95{,}833}$	$\frac{11{,}290}{95{,}833}$	$\frac{9{,}161}{95{,}833}$
Percent	19.35	59.31	11.78	9.56

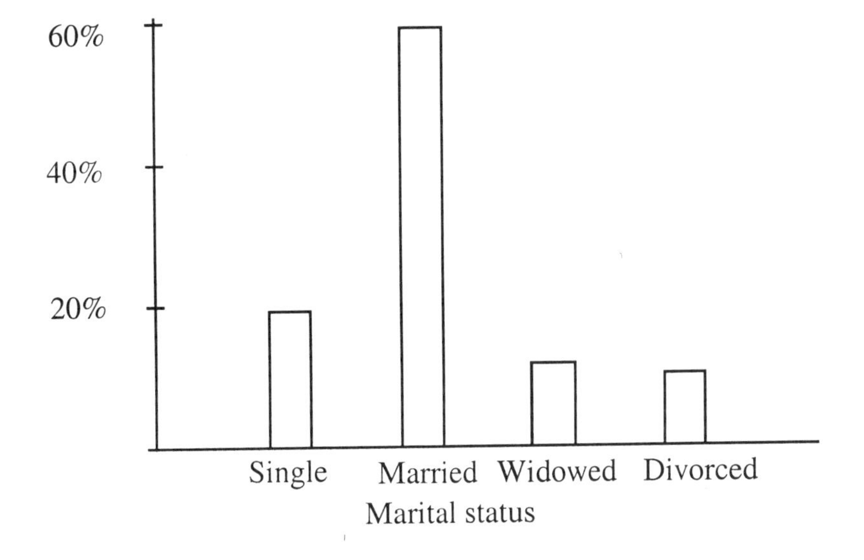

c) We give the conditional distributions, in percents, of marital status for both age groups. These percents were obtained by dividing the entry in each row by the total for the row and converting this fraction to a percent.

	Single	Married	Widowed	Divorced
18–24	71.34%	26.55%	0.06%	2.04%
40–64	5.85%	72.45%	7.61%	14.08%

The major difference is one of experience with marriage. The majority of women 18–24 (71.24%) have never been married, and few are widowed or divorced. The great majority of women 40–64 (94.14%) are or have been married. These differences probably reflect the fact that marriage, widowhood, and divorce are more likely for the older groups.

d) The distribution of ages among single women (in percents) is obtained by dividing the entries in the "Never married" column by the total for that column (seen in the table to be 19,312). The result is

	18–24	25–39	40–64	≥65
Percents	48.58%	35.91%	10.65%	4.85%

Obviously, the percent of single women decreases as age increases, and it makes sense to target women under 40. Of the women who have never been married, 84.49% are below age 40.

CHAPTER 5

PRODUCING DATA

SECTION 5.1

OVERVIEW

Data can be produced in a variety of ways. **Sampling**, when done properly, can yield reliable information about a **population**. However, even when done properly, samples are generally not appropriate for investigating cause-and-effect relations between variables. **Experiments** are investigations in which data are generated by active imposition of some treatment on the subjects of the experiment. Well-designed experiments are the best way to investigate cause-and-effect relations between variables, because they eliminate the possibility of two explanatory or lurking variables being **confounded**.

The **population** is the entire group of individuals or objects about which we want information. The information collected is contained in a **sample** that is the part of the population we actually observe. How the sample is chosen (the sampling **design**) has a large impact on the usefulness of the data. A useful sample will be representative of the **population** and will help answer our questions. "Good" methods of collecting a sample include the following:

- **Simple random samples,** sometimes denoted **SRS**
- **Probability samples**
- **Stratified random samples**
- **Multistage samples**

All these sampling methods involve some aspect of randomness through the use of a formal chance mechanism. Random selection is just one precaution that a person can take to reduce **bias,** the systematic favoring of a certain outcome. Samples we select using our own judgment, because they are convenient, or "without forethought" (mistaking this for randomness) are usually biased. This is why we use computers or a **table of random digits** to help us select a sample.

There are many ways to choose a simple random sample. A simple random sample, or SRS, of size *n* is a collection of *n* individuals chosen from the population in a manner so that each possible set of *n* individuals has an equal chance of being selected. In practice, simple methods such as drawing names from a hat is one way of getting an SRS. The "names in the hat" are the units in the population. To choose the sample, we mix up the "names" and select the sample of *n* "at random" from the hat. In reality, the population may be very large. A computer or a table of random numbers can be used to "mimic" the process of "pulling the names from the hat."

The method of selecting a SRS using a table of random digits can be summed up in two steps:

1. Give every individual in the population its own numerical label. All labels need to have the same number of digits.
2. Starting anywhere in the table (usually a spot selected at random), read off labels until you have selected as many labels as needed for the sample.

Another common type of sample design is a stratified random sample. Here, the population is first divided into **strata** and then an SRS is chosen from each strata. The strata are formed using some known characteristic of each individual thought to be associated with the response to be measured. Examples of strata are gender or age. Individuals in a particular stratum should be more like one another than those in the other strata.

Poor sample designs include the **voluntary response sample**, where people place themselves in the sample, and the **convenience sample**. Both of these methods rely on personal decision for the selection of the sample; this is generally a guarantee of bias in the selection of the sample.

Other kinds of bias can occur even in well-designed studies. Be on the lookout for:

- **Nonresponse bias**, which occurs when individuals who are selected do not participate or cannot be contacted;
- Bias in the **wording of questions** leading the answers in a certain direction;
- **Confounding** or confusing the effect of two or more variables;
- **Undercoverage** which occurs when some group in the population is given either no chance or a much smaller chance than other groups to be in the sample; and
- **Response bias**, which occurs when individuals do participate but are not responding truthfully or accurately due to the way the question is worded, the presence of an observer, fear of a negative reaction from the interviewer, or any other such source.

These types of bias can occur even in a randomly chosen sample and we need to try to reduce their impact as much as possible.

GUIDED SOLUTIONS

Exercise 5.1

KEY CONCEPTS: Populations and nonresponse

Identify the population as precisely as possible by trying to identify exactly which individuals fall in the population. Where the information is not complete, you may need to make assumptions to try to describe the population in a reasonable way. Make sure not to confuse the population of interest with the population actually sampled. There is always a strong potential for bias when they don't coincide. What is the sample?

Exercise 5.3

KEY CONCEPTS: Voluntary response sampling, bias

This is an example of a call-in opinion poll in which the subjects select themselves to be in the sample. In this case, which subjects do you think are more likely to be included in the sample, and in which direction would this bias the results?

Exercise 5.5

KEY CONCEPTS: Selecting a SRS with a table of random numbers

The table of random numbers can be used to select a SRS of numbers. In order to use it to select a random sample of six minority managers, the managers need to be assigned numerical labels. So that everyone does the problem the "same" way, we have first labeled the managers according to alphabetical order in the list.

01 - Agarwal	08 - Dewald	15 - Huang	22 - Puri
02 - Anderson	09 - Fernandez	16 - Kim	23 - Richards
03 - Baxter	10 - Fleming	17 - Liao	24 - Rodriguez
04 - Bonds	11 - Gates	18 - Mourning	25 - Santiago
05 - Bowman	12 - Goel	19 - Naber	26 - Shen
06 - Castillo	13 - Gomez	20 - Peters	27 - Vega
07 - Cross	14 - Hernandez	21 - Pliego	28 - Wang

If you go to line 139 in the table and start selecting two-digit numbers, you should get the same answer as given in the complete solution. If your entire sample has not been selected by the end of line 139, continue to the next line in the table.

The sample consists of the managers ________________________________

____.

Exercise 5.9

KEY CONCEPTS: Stratified random sample

What are the two strata from which you are going to sample? A stratified random sample consists of taking a SRS from each stratum and combining these to form the full sample. How large a SRS will be taken from each of the two strata? How would you label the units in each of the two strata from which you will sample? Fill in the following table to describe your sampling plan.

	STRATA	
	Midsize accounts	Small accounts
Number of units in stratum		
Sample size		
Labeling method		

In practice, you probably would use the table of random numbers to first select the SRS from the midsize accounts and then select the SRS from the small accounts. In this problem, you are not going to select the full samples, but only the first five units from each stratum. Start in Table B at line 115 and select 5 midsize accounts, and then continue in the table to select 5 small accounts. Write the numerical labels below.

First five midsize accounts ____________________________________.

First five small accounts ______________________________________.

Exercise 5.10

KEY CONCEPTS: Sampling frame, undercoverage

a) Which households wouldn't be in the sampling frame? Make some educated guesses as to how these households might differ from those in the sampling frame (other than the fact that they don't have phone numbers in the directory).

b) Random digit dialing makes the sampling frame larger. Which households are added to it?

Exercise 5.12

KEY CONCEPTS: Wording of questions

Questions can be worded in such a way that makes it seem as though any reasonable person should agree (disagree) with the statement. Even though both (A) and (B) address the same issue, the tone of the questions are different enough to elicit different responses. Try to decide which question should have a higher proportion favoring banning contributions.

Exercise 5.20

KEY CONCEPTS: Selecting an SRS with a table of random numbers

The table of random numbers can be used to select a SRS of blocks. In order to use it to sample blocks from the census tract, the blocks need to be assigned numerical labels. Because the blocks are already assigned three-digit numbers (the numbers do not need to be consecutive to select a SRS), enter Table B at line 125 and select three-digit numbers ignoring those three-digit numbers that don't correspond to blocks on the map. If your entire sample has not been selected by the end of line 139, continue to the next line in the table.

The sample consists of the blocks numbered ______________________________

______________ ____________________________.

Exercise 5.22

KEY CONCEPTS: Systematic sampling

a) This is like the example except there are now 200 addresses instead of 100, and the sample size is now 5 instead of 4. With these two changes, you need to think about how many different systematic samples there are. Two different systematic samples are

systematic sample 1 = 01, 41, 81, 121, 161
systematic sample 2 = 02, 42, 82, 122, 162
....

How many systematic samples are there altogether? Choosing one of these systematic samples at random is equivalent to choosing the first address in the sample. The remaining four addresses follow automatically by adding 40. Carry this out using line 120 in the table.

b) Why are all addresses equally likely to be selected? First, how many systematic samples contain each address? The chance of selecting an address is the same as the chance of selecting the systematic sample that contains it. With this in mind, what is the chance of any address being chosen? By the definition of an SRS, all samples of 5 addresses are equally likely to be selected. In a systematic sample, are all samples of 5 addresses even possible?

Exercise 5.25

KEY CONCEPTS: Inference about the population, sample size

The margin of error for all adults is three percentage points in either direction, while for men the margin of error is five percentage points in either direction. What is the effect of sample size on the accuracy of the results? How does this explain the difference in the two margins of error?

COMPLETE SOLUTIONS

Exercise 5.1

The population in the study includes all employed adult women. (The exact geographical area is not specified, but you might guess it is the U.S. It is also unclear what is meant by employed. Does this refer only to full-time employment, or are adult women working part-time to be included in the population as well?) It is important to be very clear about the population of interest before beginning a study.

The sample is obtained from a subset of this population. It is selected from the 520 members of the local business and professional women's club. These are the only people who could be selected for our sample. Any systematic differences in the opinions of the women in the population of interest and the women in the local business and professional women's club will bias the results.

The sample is the 48 questionnaires returned.

Exercise 5.3

A voluntary response sample is almost guaranteed to yield a sample that doesn't represent the entire population. A voluntary sample typically will display bias, or systematic error, in favoring some parts of the population over others. Often it is possible to use "common sense" to figure out the likely direction of bias. Are those subjects who favor tougher gun laws more likely to call? This is possible because they might be the ones dissatisfied with current laws and more likely to want to make their opinions known.

Note that the direction of bias is not as clear-cut here as in some other examples. It might have been your "educated guess" that those who like the current legislation would be more likely to call in, because they don't want the law changed, which would cause bias in the other direction. There is nothing wrong with having this opinion. The main thing is to realize that bias is likely present in this example and the reason it would occur.

Exercise 5.5

To choose a SRS of 6 managers to be interviewed, first label the members of the population by associating a two-digit number with each.

01 - Agarwal
02 - Anderson
03 - Baxter
04 - Bonds
05 - Bowman
06 - Castillo
07 - Cross
08 - Dewald
09 - Fernandez
10 - Fleming
11 - Gates
12 - Goel
13 - Gomez
14 - Hernandez
15 - Huang
16 - Kim
17 - Liao
18 - Mourning
19 - Naber
20 - Peters
21 - Pliego
22 - Puri
23 - Richards
24 - Rodriguez
25 - Santiago
26 - Shen
27 - Vega
28 - Wang

Next enter Table B and read two-digit groups until you have chosen 6 managers. Starting at line 139

55|58|8 9|94|04| 70|70|8 4|10|98 43|56|3 5|69|34| 48|39|4 5|17|19|
12|97|5 1|32|58| 13|04|8

The selected sample is 04 - Bonds, 10 - Fleming, 17 - Liao, 19 - Naber, 12 - Goel, and 13 - Gomez.

Exercise 5.9

There are 500 midsize accounts. We are going to sample 5% of these which is 25. You should label the accounts 001, 002, ..., 500, and select an SRS of 25 of the midsize accounts. There are 4400 small accounts. We are going to sample 1% of these, which is 44. You should label the accounts 0001, 0002, ..., 4400, and select an SRS of 44 of the small accounts.

Starting at line 115, we first select 5 midsize accounts, that is a SRS of size 5 using the labels 001 through 500. Continuing in the table, we select 5 small accounts, that is, an SRS of size 5 using the labels 0001 through 4400. Note that for the midsize accounts we read from Table B using three-digit numbers, and for the small accounts we read from the Table using four-digit numbers.

610|41 7|768|4 94|322| 247|09 7|3698| 1452|6 318|9332|592
1|4459| 2605|6 314|24 80|371 6|

The first five midsize accounts are those with labels 417, 494, 322, 247, and 097. Continuing in the table, using four digits instead of three, the first five small accounts are those with labels 3698, 1452, 2605, 2480, and 3716.

Exercise 5.10

a) Households omitted from the frame are those that do not have telephone numbers listed in the telephone directory. The types of people who might be underrepresented are poorer (including homeless) people who cannot afford to have a phone and the group of people who have unlisted numbers. It is harder to characterize this second group. As a group they would tend to have more money because the phone company charges to have an unlisted phone number, or it might include more single women who do not want their phone numbers available. The group might also include people whose jobs put them in contact with large groups of individuals who might harass them if their phone numbers were easily accessible.

b) People with unlisted numbers will be included in the sampling frame. The sampling frame would now include any household with a phone. One interesting point is that all households will not have the same probability of getting in the sample because some households have multiple phone lines and will be more likely to get in the sample. So, strictly speaking, random-digit dialing will not actually provide an SRS of households with phones, just an SRS of phone numbers!

Exercise 5.12

As with many pairs of questions, changes in wording of the "same" question can elicit different responses.

(A) The wording implies that contributions from special-interest groups are bad. Huge sums of money will exchange hands for the benefit of the special-interest groups.

(B) The wording suggests that we would be undemocratic to deny special-interest groups the same right to contribute money as private individuals and other groups.

We would guess that 80% favored banning contributions when presented with question (A) and 40% favored banning contributions when presented with question (B).

Exercise 5.20

The random number table starting at line 125 is reproduced below. Using three-digit numbers, select only those corresponding to blocks in the tract. The five underlined blocks — 214, 313, 409, 306, and 511 — correspond to the sample.

967|46 1|214|9 37|823| 718|68 1|844|2 35|119| 621|03 3|024|4
96|927| 199|31 3|680|9 74|192| 775|67 8|874|1 48|409| 419|03
4|390|9 99|477| 253|30 6|435|9 40|085| 169|25 8|511|7 36|071

Exercise 5.22

a) We want to select 5 addresses out of 200, so we think of the 200 addresses as five lists of 40 addresses. We choose one address from the first 40, and then every 40th address after that. The first step is to go to Table B, line 120 and choose the first two-digit random number you encounter that is one of the numbers 01, ..., 40.

35476

The selected number is 35, so the sample includes addresses numbered 35, 75, 115, 155, and 195.

b) Each individual is in exactly one systematic sample, and the systematic samples are equally likely to be chosen. In our previous example, there were 40 systematic samples, each containing 5 addresses. The chance of selecting any address is the chance of picking the systematic sample it is in, which is 1 in 40.

A simple random sample of size n would allow every set of n individuals an equal chance of being selected. Thus, in this exercise, using an SRS, the sample consisting of the addresses numbered 1, 2, 3, 4, and 5 would have the same probability of being selected as any other set of 5 addresses. For a systematically selected sample, all samples of size n do not have the same probability of being selected. In our exercise, the sample consisting of the addresses numbered 1, 2, 3, 4, and 5 would have zero chance of being selected because the numbers of the addresses do not all differ by 40. The sample we selected in part a — 35, 75, 115, 155, and 195 — had a 1 in 40 chance of being selected, so all samples of 5 addresses are not equally likely.

Exercise 5.25

Larger random samples give more accurate results than smaller sample sizes, so results based on larger samples will have smaller margins of error. Because the men are about only one-third of the sample of all adults, we will know less about the subgroup of men than we do about the group of all adults (a sample size of 472 versus a sample size of 1025 + 472 = 1397).

SECTION 5.2

OVERVIEW

An **observational study** observes individuals and measures variables of interest but does not attempt to influence the responses. On the other hand, **experiments** are studies in which one or more **treatments** are imposed on experimental **units** or **subjects**. A treatment is a specific experimental condition that is often a combination of **levels** of the explanatory variables, known as **factors**. The design of an experiment is a specification of the treatments to be used and the manner in which units or subjects are assigned to these treatments. The basic features of well-designed experiments are **control**, **randomization**, and **replication**.

Control is used to avoid confounding (mixing up) the effects of treatments with other influences, such as lurking variables. One such lurking variable is the **placebo effect**, which is the response of a subject to the fact of receiving any treatment. The simplest form of control is **randomized comparative experimentation**, which involves comparisons between two or more treatments. One of these treatments may be a **placebo** (fake treatment), and those subjects receiving the placebo are referred to as a **control group**.

Randomization can be carried out using the ideas we learned in Section 1. Randomization is carried out before applying the treatments and helps control bias by creating treatment groups that are similar. Replication, the use of many units in an experiment, is important because it reduces the chance variation between treatment groups arising from randomization. An observed effect too large to attribute plausibly to chance is called **statistically significant**. Using more units helps increases the ability of your experiment to establish differences between treatments.

Further control in an experiment can be achieved by forming experimental units into **blocks** that are similar in some way and are thought to affect the response, similar to strata in a stratified sample design. In a **block design**, units are first formed into blocks and then randomization is carried out separately in each block. **Matched pairs** are a simple form of blocking used to compare two treatments. In a matched-pairs experiment, either the same unit (the block) receives both treatments in a random order or very similar units are matched in pairs (the blocks). In the latter case, one member of the pair receives one of the treatments and the other member receives the remaining treatment. Members of a matched pair are assigned to treatments using randomization.

Some additional problems that can occur that are unique to experimental designs are lack of **blinding** and **lack of realism**. These problems should be addressed when designing the experiment.

GUIDED SOLUTIONS

Exercise 5.30

KEY CONCEPTS: Lurking variables, randomized comparative experiments

a) Are there any variables that could be different between the surgical and non-surgical treatment groups and also be related to the increased death rate? These would be lurking variables.

b) Using the examples in the textbook, construct a diagram to outline a completely randomized design for the study, using labels in your diagram that are appropriate for this study. Indicate the size of the treatment groups and a possible response variable.

Exercise 5.38

KEY CONCEPTS: Double blind experiments, bias

What does it mean for the ratings to be blind? Was this done in this case? If not, how can this bias the results and in which direction?

Exercise 5.42

KEY CONCEPTS: Matched pairs design, randomization

The first thing you should do is identify the treatments and the response variable. Next, decide what are the matched pairs in this experiment. How will you use a coin flip to assign members of a pair to the treatments? What will you measure, and how will you decide whether the right hand tends to be stronger in right-handed people?

Exercise 5.46

KEY CONCEPTS: Identifying experimental units or subjects, factors, treatments, and response variables, completely randomized design, randomization

a) You need to read the description of the study carefully. To identify the subjects, ask yourself exactly who was used in the study?

b) To identify the factor and its levels, ask yourself what question did the experiment wish to answer? What did they vary to answer this, and at how many levels?

Factor______________________________________

Levels__________________________________

c) What is measured on each subject? This is the response variable.

Response variable____________________________

Exercise 5.49

KEY CONCEPTS: Design of an experiment, randomization

a) Because the hypothesis is that calcium will reduce blood pressure, you should measure all subjects' blood pressure before starting the experiment. You must also decide on a suitable amount of time to run the experiment. It should be sufficiently long to allow the treatment to show some effect. After the allotted time has elapsed, the blood pressure of all subjects should be measured again. How do you know a reduction in blood pressure is not due to the placebo effect? Any potential problem caused by the placebo effect should be taken care of in the design. Design an appropriate experiment on these 40 men that takes the placebo effect into account. You can outline your design in words or with a picture.

b) Label the 40 names using two-digit labels. To have your answer agree with the complete solution, start with the label 01 and label down the columns. (Of course, one could start with another number or label across rows if one wished.) We need to select an SRS of 20 subjects to receive the high-calcium diet with the remainder to receive the placebo. The names are:

Alomar	Denman	Han	Liang	Rosen
Asihiro	Durr	Howard	Maldonado	Solomon
Bennett	Edwards	Hruska	Marsden	Tompkins
Bikalis	Farouk	Imrani	Moore	Townsend
Chen	Fratianna	James	O'Brian	Tullock
Clemente	George	Kaplan	Ogle	Underwood
Cranston	Green	Krushchev	Plochman	Willis
Curtis	Guillen	Lawless	Rodriguez	Zhang

Next start reading line 119 in Table B. You will need to keep reading until you have selected all the names for those in the group who will receive calcium supplements. This may require you to continue on to line 120, line 121, and subsequent lines. Because there are only two groups in this experiment, you can stop after you have selected the names for those to receive the calcium supplements. The remaining names are assigned to receive the placebo.

The subjects assigned to receive calcium supplements are ________________

____

____

____.

Exercise 5.51

KEY CONCEPTS: Placebo effect

What is the placebo effect? Does it imply that there was no physical basis for the patients pain, or does it refer to something else?

Exercise 5.53

KEY CONCEPTS: Block design, sample size

a) This is an example of a block design. Remember, a block is a group of experimental units known before the experiment to be similar in some way that is expected to affect the response to the treatments. What are the blocks in this experiment? The random assignment of units to treatments is carried out separately within each block. Your graphical representation of the experiment should illustrate the blocks and the randomization.

b) Reread the end of section 5.1 that discusses larger sample sizes. How do these ideas apply here?

COMPLETE SOLUTIONS

Exercise 5.30

a) The more seriously ill patients may be assigned to the new method by their doctors. This might be done because experience tells the doctors there is little chance of surgery being successful in these more serious cases, and there is little harm in trying the new treatment on these subjects. The seriousness of the illness would be a lurking variable and it would make the new treatment look worse than it really is.
b)

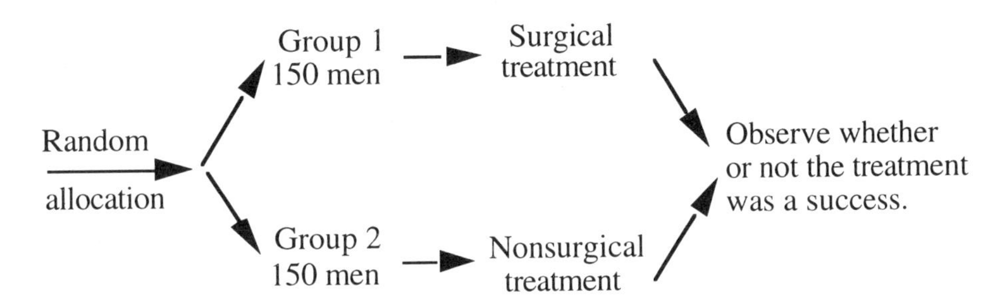

A success for the response could be no detectable cancer or alive after 8 years.

Exercise 5.38

The ratings were not blind because the experimenter who rated their level of anxiety presumably knew whether or not they were in the meditation group. Because the experimenter was hoping to show that those that meditated had lower levels of anxiety, he might unintentionally rate those in the meditation group as having lower levels of anxiety, if there is any subjectivity in the rating. This would make the meditation look more effective than it really is. It would be better if a third party who did not know which group the subjects belonged to rated the subjects anxiety levels.

Exercise 5.42

We have 10 subjects available. There are two treatments in the study. Treatment 1 is squeezing with the right hand and treatment 2 is squeezing with the left hand. The response is the force exerted as indicated by the reading on the scale.

To do the experiment, we use a matched-pairs design. The matched pairs are the two hands of a particular subject. We should randomly decide which hand to use first, perhaps by flipping a coin. We measure the response for each hand and then compare the forces for the left and right hands over all subjects to see whether there is a systematic difference between the two hands.

Exercise 5.46

a) The subjects are the 210 children aged 4 to 12 years who participated in the study. The problem tells us nothing about how these children were selected to be in the study.

b) The factor is the set of choices that are presented to each subject. The levels correspond to the three sets of choices, Set 1 (2 milk, 2 fruit), Set 2 (4 milk, 2 fruit), and Set 3 (2 milk, 4 fruit), so there are three levels in this problem. (You might have thought the levels were the number of choices presented to the subject, but Sets 2 and 3 each has six choices and correspond to different levels of the factor.)

c) The response variable is whether they chose a milk drink or a fruit drink. Since they had to choose one or the other, the response just could be whether or not they chose a milk drink (yes or no).

Exercise 5.49

a) Measure all subjects' blood pressure before starting the experiment. Assign 20 subjects at random to the added calcium diet and the remainder to the placebo group. After a suitable amount of time has passed for the treatment to show some effect, measure the blood pressure again and compare the change for the two groups. If the calcium is effective, the calcium group should show a *greater* lowering of blood pressure than the placebo group. (Remember that the placebo group might also show some lowering of their blood pressure. That's the placebo effect.) In a diagram, the design is represented as

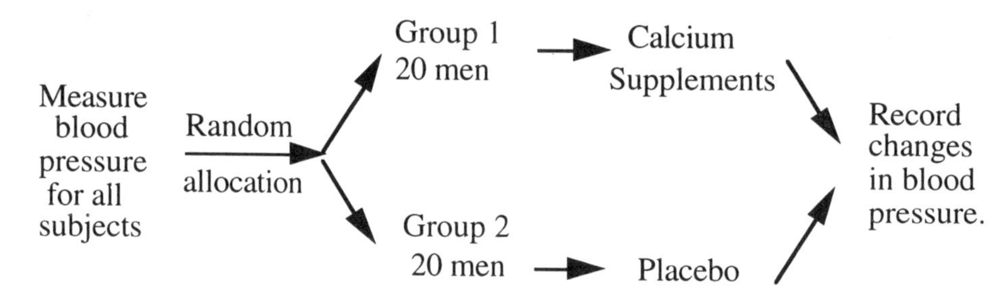

b) The names with their labels are

01 Alomar	09 Denman	17 Han	25 Liang	33 Rosen
02 Asihiro	10 Durr	18 Howard	26 Maldonado	34 Solomon
03 Bennett	11 Edwards	19 Hruska	27 Marsden	35 Tompkins
04 Bikalis	12 Farouk	20 Imrani	28 Moore	36 Townsend
05 Chen	13 Fratianna	21 James	29 O'Brian	37 Tullock
06 Clemente	14 George	22 Kaplan	30 Ogle	38 Underwood
07 Cranston	15 Green	23 Krushchev	31 Plochman	39 Willis
08 Curtis	16 Guillen	24 Lawless	32 Rodriguez	40 Zhang

Next, start reading line 119 in Table B. Read across the row in groups of digits equal to the number of digits you used for your labels. (Because you used two digits for labels, read line 119 in pairs of digits.)

95857 07118 87664 92099 58806 66979 98624 84826
35476 55972 39421 65850 04266 35435 43742 11937
71487 09984 29077 14863 61683 47052 62224 51025
13873 81598 95052 90908 73592 75186 87136 95761

Those assigned to the calcium supplemented diet are 18 – Howard, 20 – Imrani, 26 – Maldonado, 35 –Tompkins, 39 – Willis, 16 – Guillen, 04 – Bikalis, (26 and 35 are the next two numbers but both are skipped because they have been selected already), 21 – James, 19 – Hruska, 37 – Tullock, 29 – O'Brian, 07 – Cranston, 34 – Solomon, 22 – Kaplan, 10 – Durr, 25 – Liang, 13 – Fratianna, 38 – Underwood, 15 – Green, and 05 – Chen. The remaining men are in the placebo group.

Exercise 5.51

Many patients respond favorably to any treatment, even a placebo that has no therapeutic effect. There could have been a physical basis for the patient's pain and they are responding to the fact of having received *any* treatment, which is the meaning of the placebo effect.

Exercise 5.53

a)

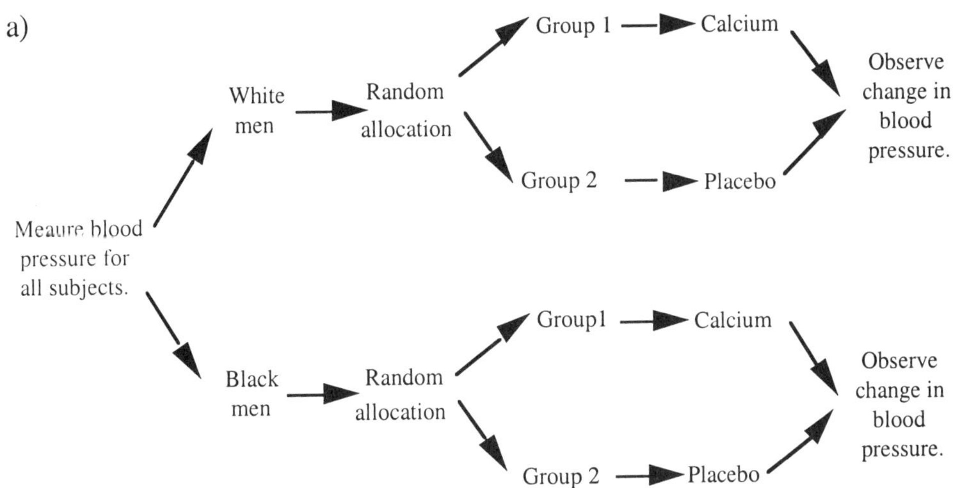

b) A larger group of subjects will provide more information because individual differences tend to be more evenly balanced between the treatments (all those with high blood pressure are less likely to be in the same group). When comparing sample averages between the treatments, these averages will be more likely to reflect treatment differences and be less influenced by individual differences.

SECTION 5.3

OVERVIEW

Complex probabilities for many random phenomena can be approximated using a carefully designed **simulation**. This requires formulating a model for the phenomenon from which we can use random digits from either a table, graphing calculator, or computer software to simulate many repetitions quickly. Because the proportion of repetitions on which a particular result occurs will be close to its true probability for a large number of repetitions, simulation can be used to get good estimates of probabilities. The steps involved in carrying out a simulation are

Step 1: State the problem or describe the experiment.
Step 2: State the assumptions.
Step 3: Assign digits to represent the outcomes.
Step 4: Simulate many repetitions.
Step 5: Calculate relative frequencies and give conclusions.

GUIDED SOLUTIONS

Exercise 5.55

KEY CONCEPTS: Simulation, table of random digits

a) First, you must decide which digits you will use for the Democrats and which you will use for the Republicans. There are many choices here. Because Americans are equally divided on whether Democrats or Republicans are better able to manage the economy (50% think Democrats and 50% think Republicans), you must use the same number of digits for each. For example, you could let Americans choosing Democrats correspond to odd digits and Americans choosing Republicans correspond to even digits, or you could let Democrats correspond to the digits 0, 1, 2, 3, 4 and Republicans correspond to the digits 5, 6, 7, 8, 9. Going to line 110 in Table B and using the first choice for assignment of digits gives the following simulation of 10 independently chosen adults:

3 8 4 4 8 4 8 7 8 9
DRRRR RRDRD

This gives 3 Democrats and 7 Republicans.

b) What digits would you assign to each Democrat and Republican? How would you allow for the different proportion of adults giving each choice? Next, go to line 111 and simulate the responses of 10 independently chosen adults.

c) How would you allow for the three possible responses to the question? What digits would you assign to each response? Next, go to line 112 and simulate the responses of 10 independently chosen adults.

d) How many random digits would you need to simulate the response of each voter for this experiment? Next, go to line 113 and simulate the responses of 10 independently chosen adults.

Exercise 5.63

KEY CONCEPTS: Simulation, table of random digits

a) There are 52 cards in the deck and each gets a two-digit random number. For convenience, we can let the pairs of digits 01, 02, ..., 13 correspond to the 13 hearts and the digits 14, 15, ..., 52 correspond to the other 39 cards. Ignore all other two-digit numbers. To carry out the simulation of drawing *two* cards, how many digits are needed in total? What would you do if you got the sequence of digits 0101?

b) Using a calculator, the table of random digits or computer software, carry out 25 rounds of this game. In the complete solutions, the 25 rounds of the game are simulated starting at line 130 in Table B. The actual probability of winning $1 can be worked out using more advanced methods in probability and is 0.4412.

COMPLETE SOLUTIONS

Exercise 5.55

a) Answer given in the guided solutions.

b) One choice of digits is to let Americans choosing Democrats correspond to the digits 0, 1, 2, 3, 4, 5 and Americans choosing Republicans correspond to the digits 6, 7, 8, 9. With this choice of digits, the simulated responses of 10 independently chosen adults starting in line 111 is

8 1 4 8 6   6 9 4 8 7
RDDRR   RRDRR

giving 3 Democrats and 7 Republicans.

c) Because there are now three outcomes, we must assign random digits to each. The following choice is one of many possibilities.

Democrat (D)	0, 1, 2, 3
Republican (R)	4, 5, 6, 7
Undecided (U)	8, 9

With this choice of digits, the simulated responses of 10 independently chosen adults starting in line 112 is

5 9 6 3 6   8 8 8 0 4
RURDR   UUUDR

giving 2 Democrats, 3 Republicans, and 4 Undecided.

d) We need two random digits to simulate the response of each voter for this experiment. The following choice is one of many possibilities.

Democrats (D)	00, 01, ..., 52
Republicans (R)	53, 54, ..., 99

With this choice of digits, the simulated responses of 10 independently chosen adults starting in line 113 is

62|56|8 7|02|06| 40|32|5 0|36|99
R R R D D D D D D R

giving 6 Democrats and 4 Republicans.

Exercise 5.63

a) There are 52 cards in the deck and each gets a two-digit random number. For convenience, we can let the pairs of digits 01, 02, ..., 13 correspond to the 13 hearts and the digits 14, 15, ..., 52 correspond to the other 39 cards. Ignore all other two-digit numbers. To simulate a round of drawing two cards, you need two distinct two-digit numbers. If you get the same two-digit number twice in a row (it will occasionally happen and would correspond to getting the same card on both draws which is impossible), then ignore the second two-digit number and let the second card correspond to the next two digits in the table.

b) If you were using a Table B, line 130, the simulation would start as follows.

69|05|1 6|48|17| 87|17|4 0|95|17|

Skip the 69 because it doesn't corespond to one of the 52 cards in our assignment of digits. The 05 corresponds to a heart and the 16 corresponds to a non-heart. Because you have at least one heart, you win $1. Reshuffling is equivalent to continuing in the table. The 48 corresponds to a non-heart and the 17 corresponds to a non-heart. Because neither of these two cards is a heart, you would lose $1. Continuing in Table B, the following are the 25 simulations, where W corresponds to winning $1 and L corresponds to losing $1.

WLLLL WLLWL LLLLL WWWWW LWLLL

This gives 9 wins on 25 rounds or an estimated probability of winning using the simulation of 9/25 = 0.36. (Note: Because it is possible to calculate the probability of winning exactly as 0.4412 using more advanced methods in probability, we wouldn't ordinarily use a simulation for this experiment. In practice, simulations are used when the answers are too complicated for exact calculations.)

SELECTED TEXT REVIEW EXERCISES

GUIDED SOLUTIONS

Exercise 5.66

KEY CONCEPTS: Populations and samples, sample size, bias

a) Try to identify the population as exactly as possible. Where the information is incomplete, you may need to make assumptions to try to describe the population in a reasonable way. What is the sample in this example?

b) Are the sample sizes large enough to have fairly accurate information about the population? Is there a possibility of response bias in this example?

Exercise 5.68

KEY CONCEPTS: Observational studies and experiments

What are the groups or "treatments" in this study? How were the subjects assigned to the groups? Was a treatment deliberately imposed on individuals to observe their responses? It may help to think about what the explanatory and response variables are in this example.

Exercise 5.72

KEY CONCEPTS: Explanatory and response variables, factors and levels, completely randomized design

a) The individuals on which the experiment is done are the experimental units. What are the experimental units in this study? What response is being observed?

Experimental units =

Response =

b) The explanatory variables in an experiment are the factors. How many factors are there in this experiment? When there is more than one factor in an experiment, a "treatment" is formed by combining a level of each of the factors. How many levels of each factor are there, and how many treatments can be formed? If there are 10 chicks for each treatment, how many experimental units are needed. Use a diagram to illustrate the different combinations of the levels of the factors to form the treatments.

c) The experiment is run as a completely randomized design with how many treatments? Use a diagram to illustrate the design for this experiment.

COMPLETE SOLUTIONS

Exercise 5.66

a) The population is Ontario residents. From the problem, there is no reason to believe that the population is restricted to adult residents or any other subgroup. Any person residing in Ontario is in the population. The sample is the 61,239 people who were interviewed.

b) The sample size is quite large, and in an SRS you would suspect about half the sample to be men and half to be women, so there should be a reasonably large sample size from each of the groups. This should give you estimates that are pretty close to the truth about the entire population unless there is bias present.

This example is very similar to the example used in the discussion of response bias. Answers to questions that ask respondents to recall past events are often inaccurate because of faulty memory. People may bring past events forward to more recent time periods, so that someone who had visited a general practitioner 14 months ago might answer yes to this question. These estimates might be larger than the true percentage due to response bias. Good interviewing techniques can reduce this source of bias.

Exercise 5.68

The subjects are all executives who have volunteered for an exercise program. Each receives a physical examination, which is used to divide them into two groups corresponding to low fitness and high fitness. The experimenter did not impose the level of physical fitness on the subjects. Their level of physical fitness was already present, which placed them in the groups, not the experimenter. The experimenter had no role in who was assigned to low fitness and high fitness. Those who are in the low fitness group may be heavier, include executives with significant health problems, or include executives who are adverse to a disciplined lifestyle, and these could be a lurking variables that might explain differences in leadership.

This is an observational study. The individuals were observed and the variables of interest, physical fitness, and leadership were measured. There was no attempt to influence the response variable leadership by altering the subjects' level of physical fitness.

Exercise 5.72

a) The 1-day-old male chicks are the experimental units. They are the objects that receive the treatment. The response variable is the amount of weight they gain in 21 days, or the difference between their initial weight and their weight after three weeks.

b) This is an example of an experiment with two factors. The first factor is the variety of corn, which has three levels: normal, opaque-2, and floury-2. The second factor is the protein level of the diet, which has three levels: 12% protein, 16% protein, and 20% protein. A diet or treatment is made up of a combination of a type of corn and a level of protein so there are $3 \times 3 = 9$ possible treatments or diets. Because 10 chicks are assigned to each diet, the experiment needs $9 \times 10 = 90$ experimental units. The figure below shows the layout of the treatments.

		Factor B Variety		
		Normal	Opaque-2	Floury-2
Factor A % protein	12%	1	2	3
	16%	4	5	6
	20%	7	8	9

Combinations of the levels of the two factors form nine treatments

c) In the diagram on the next page, Group 1 gets treatment 1, group 2 gets treatment 2, etc., where the treatment numbers correspond to those in part b. There are nine treatments and nine groups.

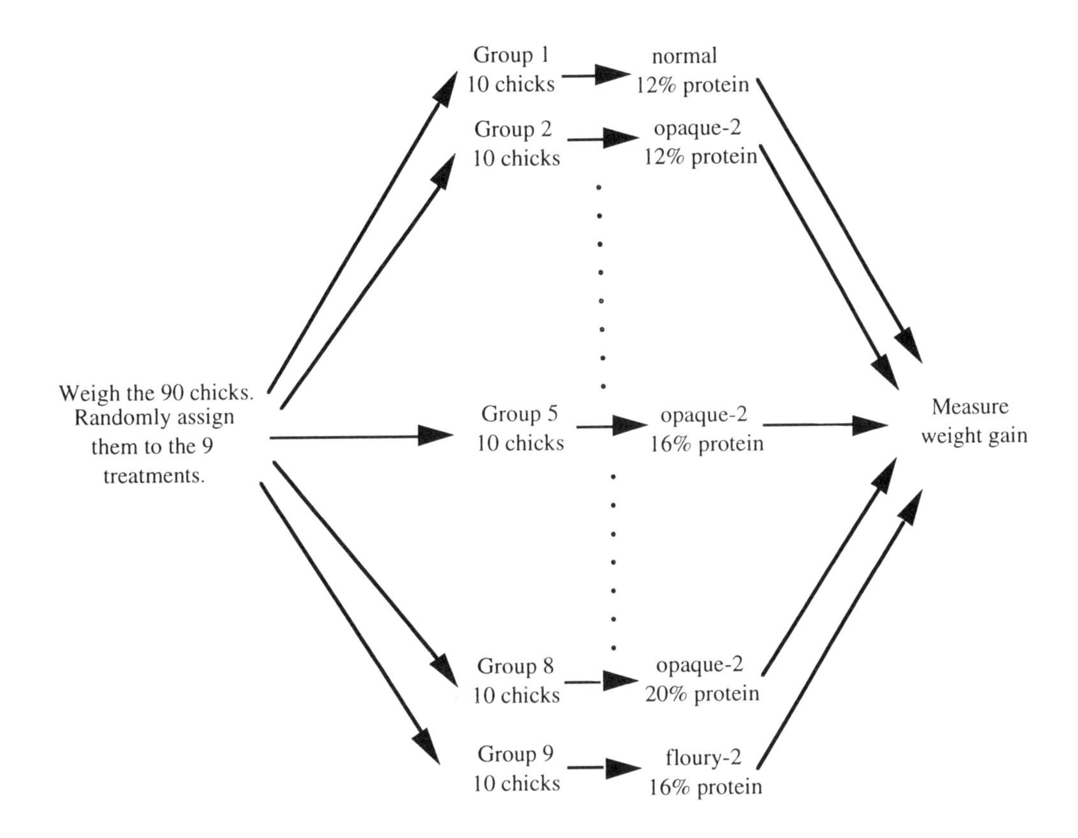

Weigh the 90 chicks. Randomly assign them to the 9 treatments.
Group 1
10 chicks
normal
12% protein
Group 2
10 chicks
opaque-2
12% protein
Group 5
10 chicks
opaque-2
16% protein
Group 8
10 chicks
opaque-2
20% protein
Group 9
10 chicks
floury-2
16% protein
Measure weight gain

CHAPTER 6

PROBABILITY: THE STUDY OF RANDOMNESS

SECTION 6.1

OVERVIEW

A process or phenomenon is called **random** if its outcome is uncertain. Although individual outcomes are uncertain, the underlying distribution for the possible outcomes begins to emerge when the process is repeated a large number of times. For any outcome, its **probability** is the proportion of times, or the relative frequency, with which the outcome would occur in a long series of repetitions of the process. It is important that these repetitions or trials be **independent** for this property to hold.

You can study random behavior by carrying out physical experiments such as coin tossing or rolling of a die, or you can simulate a random phenomenon on a calculator or a computer. Using the computer is particularly helpful when we want to consider a large number of trials.

GUIDED SOLUTIONS

Exercise 6.6

KEY CONCEPTS: Assigning probabilities

This is a good "game" to play to learn about random behavior, and it's an exercise for which everyone will get different answers. What you need to do is list the 20 outcomes in terms of Betty's possible winnings (–$4, –$2, $0, $2, $4) and complete the relative frequency table below. These relative frequencies represent guesses or estimates of the true probabilities of the different outcomes. Because the relative frequencies are based on only 20 trials, it's likely that they will not be that close to the true probabilities, but you should be able to see which outcomes are more or less likely.

Outcome	Relative frequency
–$4	
–$2	
$0	
$2	
$4	

Exercise 6.7

KEY CONCEPTS: Simulating a random phenomenon

a) You will need to use your calculator (or computer software) to simulate the 100 trials. If you are using a TI-83, follow the directions given in the statement of the problem in the textbook.

b) Proportion of hits =

b) You need to go through your sequence to determine the longest string of hits or misses.

Longest run of shots hit =   Longest run of shots missed =

COMPLETE SOLUTIONS

Exercise 6.6

Our 20 trials yielded the following outcomes.

0	0	0	2	0	0	-4	0	2	2
-2	2	2	-4	0	-2	0	2	2	0

The relative frequencies of the outcomes are given below. We also have computed the true probabilities (long-term relative frequencies) using methods that you will learn later in this chapter. The relative frequencies show the general pattern, but with only 20 trials, the differences between the relative frequencies and true probabilities are still quite large. The agreement would be improved if the number of trials were increased.

Outcome	Relative frequency	True Probability
–4	2/20 = .10	.0625
–2	2/20 = .10	.2500
0	9/20 = .45	.3750
2	7/20 = .35	.2500
4	0/20 = .00	.0625

Exercise 6.7

a) Our sequence of hits (H) and misses (M) is given below.

```
H  H  M  H  H  H  M  M  H  H  H  M  H  M  H
M  H  M  M  H  M  M  H  H  H  M  M  H  H  H
M  M  M  M  H  M  M  H  H  H  H  M  H  H  H
M  M  M  M  M  H  M  H  H  M  H  M  H  M  M
H  H  H  H  M  H  M  M  M  M  H  H  M  H  H
M  M  H  H  H  M  M  H  H  M  M  H  M  H  M
M  H  M  H  H  M  H  H  H  H
```

b) proportion of hits = .54

c) You need to go through your sequence to determine the longest string of hits or misses. In our example,

Longest run of shots hit = 4 (this occurred more than once)
Longest run of shots missed = 5

SECTION 6.2

OVERVIEW

The description of a random phenomenon begins with the **sample space** which is the list of all possible outcomes. A set of outcomes is called an **event.** When we have determined the sample space, a **probability model** tells us how to assign probabilities to the various events that can occur. There are four basic rules that probabilities must satisfy.

- Any probability is a number between 0 and 1.
- All possible outcomes together must have probability 1.

- The probability of the **complement** of an event (the probability that the event does not occur) is 1 minus the probability that the event occurs.
- If two events have no outcomes in common, the probability that one or the other occurs is the sum of their individual probabilities.

In a sample space with a finite number of outcomes, probabilities are assigned to the individual outcomes, and the probability of any event is the sum of the probabilities of the outcomes that it contains. In some special cases, the outcomes are all **equally likely** and the probability of any event A is just computed as

$$P(\text{A}) = (\text{number of outcomes in A})/(\text{number of outcomes in S}).$$

Events are **disjoint** if they have no outcomes in common. In this special case, the probability that one or the other event occurs is the sum of their individual probabilities. This is the **addition rule** for disjoint events, namely

$$P(\text{A or B}) = P(\text{A}) + P(\text{B}).$$

Events are **independent** if knowledge that one event has occurred does not alter the probability that the second event occurs. The mathematical definition of independence leads to the **multiplication rule** for independent events. If A and B are independent, then

$$P(\text{A and B}) = P(\text{A})P(\text{B}).$$

In any particular problem, we can use this definition to check whether two events are independent by seeing whether the probabilities multiply according to the definition. Most of the time, however, independence is assumed as part of the probability model. The four basic rules, plus the multiplication rule, allow us to compute the probabilities of events in many random phenomena.

Many students confuse independent and disjoint events once they have seen both definitions. Remember, disjoint events have no outcomes in common and when two events are disjoint, you can compute $P(\text{A or B}) = P(\text{A}) + P(\text{B})$ in this special case. The probability being computed is that one *or* the other event occurs. Disjoint events cannot be independent because once we know that A has occurred, then the probability of B occurring becomes 0 (B cannot have occurred as well — this is the meaning of disjoint). The multiplication rule can be used to compute the probability that two events occur *simultaneously*, $P(\text{A and B}) = P(\text{A})P(\text{B})$, in the special case of independence.

GUIDED SOLUTIONS

Exercise 6.10

KEY CONCEPTS: Sample space

One of the main difficulties encountered when describing the sample space is finding some notation to express your ideas formally. Following the textbook, our general format is $S = \{\quad\}$, where a description of the outcomes in the sample space is included within the braces.

a) You want to express that any number between 0 and 24 is a possible outcome. So you would write $S = \{$all numbers between 0 and 24$\}$.

b) $S =$

c) $S =$

d) You may not want to put an upper bound on the amount, so you should allow for any number greater than 0. $S =$

e) Remember that the rats can lose weight. $S =$

Exercise 6.18

KEY CONCEPTS: Applying the probability rules

a) Because these are the only blood types, what has to be true about the sum of the probabilities for the different types? Use this to find P(AB).

b) What's true about the events O and B blood type? Which probability rule do we follow? (Don't be confused by the wording in the problem, which says "people with blood types O *and* B." In the context of this problem and the language of probability we are using, it really means O or B. There are no people with blood types O *and* B.)

Exercise 6.24

KEY CONCEPTS: Independence, multiplication rule

The probability of winning the major battle is 0.6. What is the probability of winning all three small battles? How would you decide which strategy is best? Write the event "winning all three small battles" in terms of winning each of the small battles and use the independence of victories or defeats in the small battles.

P(winning all three small battles) =

Which strategy do you prefer and why?

Exercise 6.33

KEY CONCEPTS: Multiplication rule for independent events

Look at the first five rolls of each of the sequences. They have the same probability, since there is one G and 4 R's. So for the first five rolls, none of the sequences would be preferred (they all have the same probability). On the sixth roll, in the first sequence you could get either a G or an R and you would win. For the second sequence, you need a G on the sixth roll to win, and for the third sequence you need an R. So this makes the first sequence the highest probability. Why do you think most students chose the second sequence? To check, compute the probabilities for each of the three sequences using the multiplication rule.

P(First sequence) =

P(Second sequence) =

P(Third sequence) =

Exercise 6.35

KEY CONCEPTS: Multiplication rule for independent events

a) The three years are independent. If U indicates a year for the price being up and D indicates a year for the price being down, you need to compute *P*(UUU).

b) Because the events are independent, what happens in the first two years does not affect the probability of going up or down in the third year. What's the probability of the price going down in any given year?

c) This problem must be set up carefully and done in steps.

Step 1: Write the event of interest in terms of simpler outcomes. How would you write P(UU or DD) in terms of P(UU) and P(DD)?

P(moves in the same direction in the next two years) = P(UU or DD).

Step 2: Evaluate P(UU) and P(DD) and substitute your answer in the expression from step 1.

COMPLETE SOLUTIONS

Exercise 6.10

a) S = {all numbers between 0 and 24}.
b) S = {0, 1, 2, ..., 11,000}.
c) S = {0, 1, ..., 12}.
d) S = {set of all numbers greater than or equal to zero}.
e) S = {all positive and negative numbers}. The inclusion of the negative numbers allows for the possibility that the rats lose weight.

Exercise 6.18

a) The probabilities for the different blood types must add to 1. The sum of the probabilities for blood types O, A, and B is 0.49 + 0.27 + 0.20 = 0.96. Subtracting this from 1 tells us that the probability of the remaining type AB must be 0.04.

b) Maria can receive transfusions from people with blood types O or B. Because a person cannot have both of these blood types, they are disjoint. The calculation follows probability rule 4, which says P(O or B) = P(O) + P(B) = 0.49 + 0.20 = 0.69.

Exercise 6.24

Denote the event that the general wins the first small battle by W_1, the event that the general wins the second small battle by using W_2, and the event that the general wins the third small battle by using W_3. Then

P(winning all three small battles) = $P(W_1$ and W_2 and $W_3) = P(W_1)P(W_2)P(W_3)$

since victories or defeats in the small battles are independent. We know that the probability of winning each small battle is 0.8, so

$$\begin{aligned} P(\text{winning all three small battles}) &= P(W_1)P(W_2)P(W_3) = 0.8 \times 0.8 \times 0.8 \\ &= 0.512. \end{aligned}$$

Because the general is more likely to win the major battle than to win all three small battles, his strategy should be to fight one major battle.

Exercise 6.33

You need to use the multiplication rule since the rolls are independent. For the first sequence

$P(\text{RGRRR}) = \left(\frac{2}{6}\right)\left(\frac{4}{6}\right)\left(\frac{2}{6}\right)\left(\frac{2}{6}\right)\left(\frac{2}{6}\right)$. In the product for the probability of this sequence, the 2/6 appears 4 times, once for each R outcome, and the 4/6 appears once, for the G outcome.

$$P(\text{First sequence}) = \left(\frac{2}{6}\right)^4\left(\frac{4}{6}\right) = 0.0082$$

$$P(\text{Second sequence}) = \left(\frac{2}{6}\right)^4\left(\frac{4}{6}\right)^2 = 0.0055$$

$$P(\text{Third sequence}) = \left(\frac{2}{6}\right)^5\left(\frac{4}{6}\right) = 0.0027$$

Exercise 6.35

a) $P(\text{UUU}) = (0.65)^3 = 0.2746$.

b) The probability of the price being down in any given year is $1 - 0.65 = 0.35$. Because the years are independent, the probability of the price being down in the third year is 0.35, regardless of what has happened in the first two years.

c) P(moves in the same direction in the next two years) = P(UU or DD) = P(UU) + P(DD), because the events UU and DD are disjoint. Using the independence of two successive years, $P(\text{UU}) = (0.65)^2 = 0.4225$, and $P(\text{DD}) = (0.35)^2 = 0.1225$. Putting this together,

P(moves in the same direction in the next two years) = 0.4225 + 0.1225 = 0.5450.

SECTION 6.3

OVERVIEW

This section discusses a number of basic concepts and rules that are used to calculate probabilities of complex events. The **complement** A^c of an event A contains all the outcomes in the sample space that are not in A. It is the "opposite" of A. The **union** of two events A and B contains all outcomes in A, in B, or in both. The union is sometimes referred to as the event A or B. The **intersection** of two events A and B contains all outcomes that are in both A and B simultaneously. The intersection is sometimes referred to as the event A and B. We say two events A and B are **disjoint** if they have no outcomes in common.

The **conditional probability** of an event B given an event A is denoted $P(B|A)$ and is defined by

$$P(B|A) = \frac{P(A \text{ and } B)}{P(A)}$$

when $P(A) > 0$. Two events A and B are **independent** if $P(B|A) = P(B)$. In practice, conditional probabilities and independence often can be determined directly from the information given in a problem.

Other general rules of elementary probability are

- Legitimate values: $0 \le P(A) \le 1$ for any event A
- Total probability: $P(S) = 1$, where S denotes the sample space.
- Complement rule: $P(A^c) = 1 - P(A)$
- Addition rule: $P(A \text{ or } B) = P(A) + P(B) - P(A \text{ and } B)$
- Multiplication rule: $P(A \text{ and } B) = P(A)P(B|A)$
- For disjoint events: $P(A \text{ and } B) = 0$ and so $P(A \text{ or } B) = P(A) + P(B)$
- For independent events: $P(A \text{ and } B) = P(A)P(B)$

In problems with several stages, it is helpful to draw a tree diagram to guide you in the use of the multiplication and addition rules.

GUIDED SOLUTIONS

Exercise 6.37

KEY CONCEPTS: The addition rule

This is an application of the addition rule

$$P(\text{A or B}) = P(\text{A}) + P(\text{B}) - P(\text{A and B}).$$

Exercise 6.39

KEY CONCEPTS: Venn diagrams

a) From exercise 6.37, we have the following facts about A and B.

$P(\text{A}) = 0.125$
$P(\text{B}) = 0.237$
$P(\text{A and B}) = 0.077$
$P(\text{A or B}) = 0.285$
$P(\text{entire sample space}) = 1$

Below is a Venn diagram showing A and B. Shade the portion representing {A and B}.

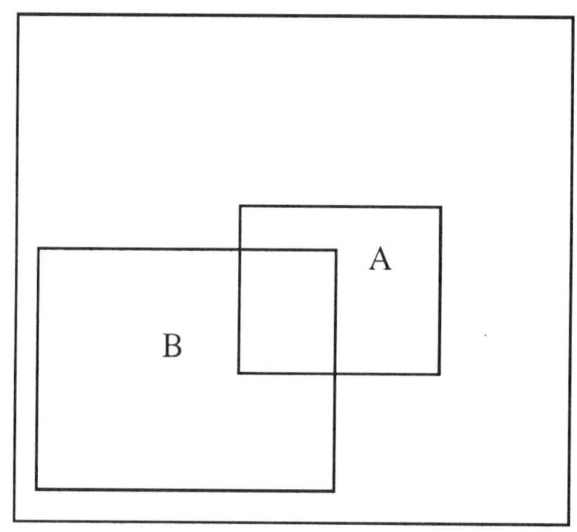

We are given that $P(\text{A and B}) = 0.077$. Next, describe in words what the event A and B means. Refer to exercise 6.37 for the meaning of A and B.

b) Below is a Venn diagram showing A and B. Shade the portion representing {A and B^c}.

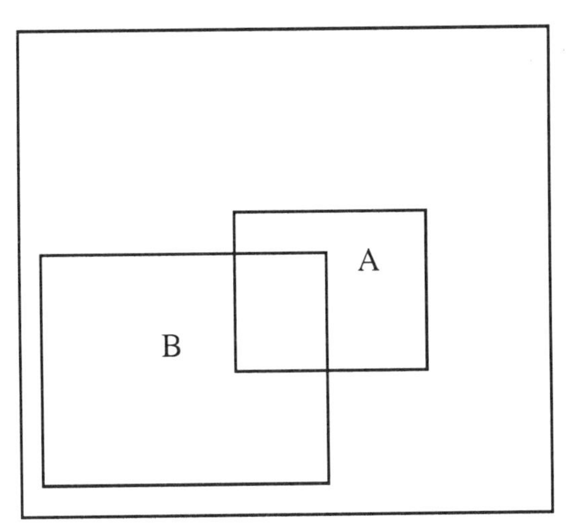

Use the facts listed in part a and the Venn diagram to calculate $P(\text{A and B}^c)$.

Next, describe in words what the event A and B^c means. Refer to exercise 6.37 for the meaning of A and B.

c) Below is a Venn diagram showing A and B. Shade the portion representing $\{A^c \text{ and } B\}$.

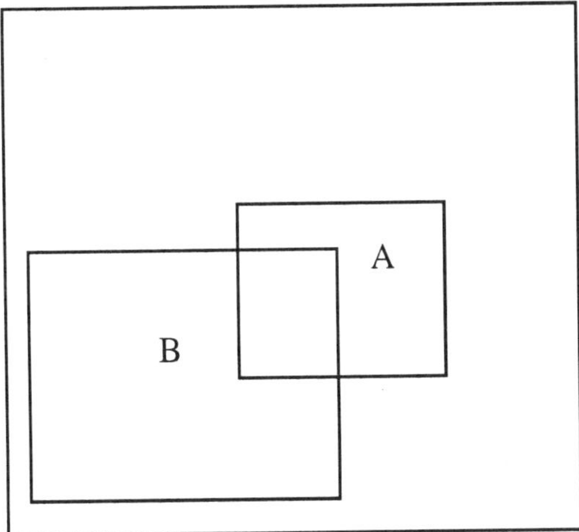

Use the facts listed in part a and the Venn diagram to calculate $P(A^c \text{ and B})$.

Next, describe in words what the event A^c and B means. Refer to exercise 6.37 for the meaning of A and B.

d) Below is a Venn diagram showing A and B. Shade the portion representing $\{A^c \text{ and } B^c\}$.

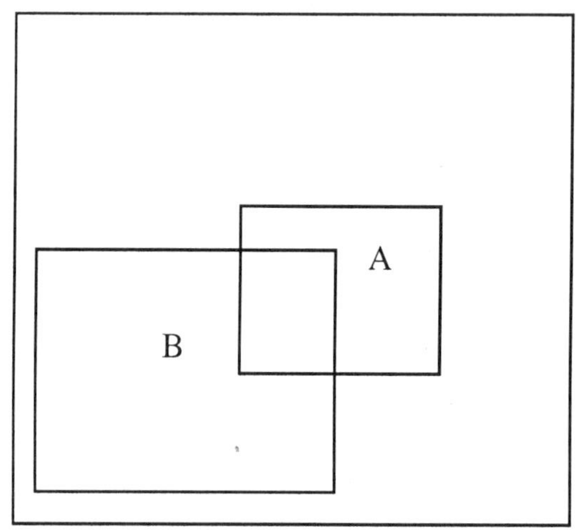

Use the facts listed in part a and the Venn diagram to calculate $P(A^c \text{ and } B^c)$.

Next, describe in words what the event A^c and B^c means. Refer to exercise 6.37 for the meaning of A and B.

Exercise 6.41

KEY CONCEPTS: Conditional probabilities and the multiplication rule

Table 6.1 is reproduced below to assist you.

	Age			
	18 to 24	25 to 64	65 and older	Total
Married	3046	48116	7767	58929
Never married	9289	9252	768	19309
Widowed	19	2425	8636	11080
Divorced	260	8916	1091	10267
Total	12614	68709	18262	99585

a) You can calculate this probability directly from the table. All women in the table are equally likely to be selected (that is what it means to select a woman at random) so that the fraction of the women in the table 65 years or older is the desired probability. How many women are 65 years or older? Where do you find this in the table? What is the total number of women represented in the table? Use these numbers to compute the desired fraction.

b) This probability also can be calculated directly from the table. Because this is a conditional probability (i.e., this is a probability given that the woman is 65 or older), we restrict ourselves only to women who are 65 or older. The desired probability is then the fraction of these women who are married. Find the appropriate entries in the table to compute this fraction.

c) The number of women who are both married and in the 65 or older age group can be read directly from the table. What is this number? What fraction of the total number of women represented in the table is this number? This is the desired probability.

d) Recall the multiplication rule says

P(woman is both "married" and "65 and older")

$=$ P(woman is married|woman is 65 and over)P(woman is 65 and older)

Next, verify this using the answers to parts a, b, and c.

Exercise 6.53

KEY CONCEPTS: Tree diagrams

We are given the following probabilities.

> *P*(Linda knows the answer) = 0.75
>
> *P*(Linda is correct | Linda does not know the answer) = 0.2

It also must be the case that

> *P*(Linda is correct | Linda knows the answer) = 1.

Next, fill in the probabilities for the branches in the tree diagram below and compute *P*(Linda is correct). Remember that the sum of the probabilities of all branches starting from the same point in the diagram is 1, assuming that all possibilities from the starting point are represented by the branches (you can verify that this is true in the tree diagram below). The probability of any endpoint in the tree is the product of the probabilities on the branches leading from the beginning of the tree to the endpoint. *P*(Linda is correct) is the sum of all the probabilities of endpoints labeled "Linda is correct."

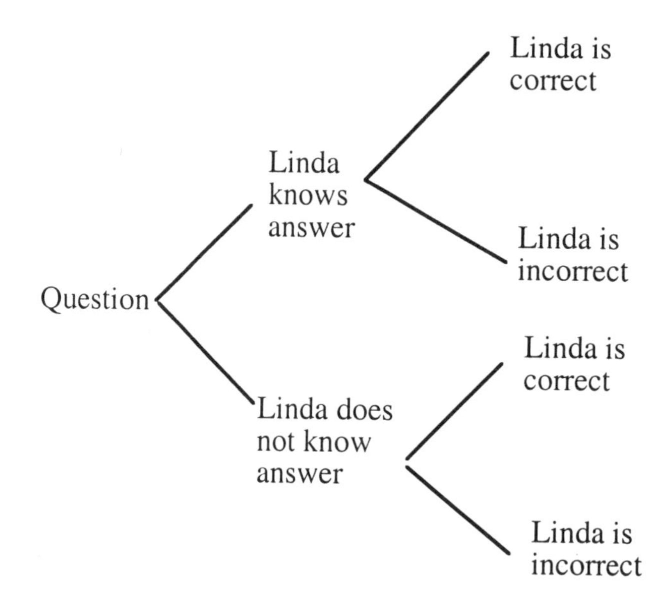

Exercise 6.55

KEY CONCEPTS: Bayes rule

What we want to calculate is

P(Linda knows answer | Linda is correct).

Note that this is not 1, because Linda might not know the answer but by good luck guesses the correct answer.

To calculate the probability, we could use the definition of conditional probability, which says

$$P(\text{B}|\text{A}) = \frac{P(\text{A and B})}{P(\text{A})}$$

For our problem, what are A and B? What are the various probabilities on the right side of this equation ? (Hint: Refer to exercise 6.53 for these probabilities. They can be determined from the tree diagram.)

COMPLETE SOLUTIONS

Exercise 6.37

We are given that P(A) = 0.125, P(B) = 0.237, and PA and B) = 0.077, hence

P(A or B) = 0.125 + 0.237 – 0.077 = 0.285

Exercise 6.39

a) The shaded area below is {A and B}

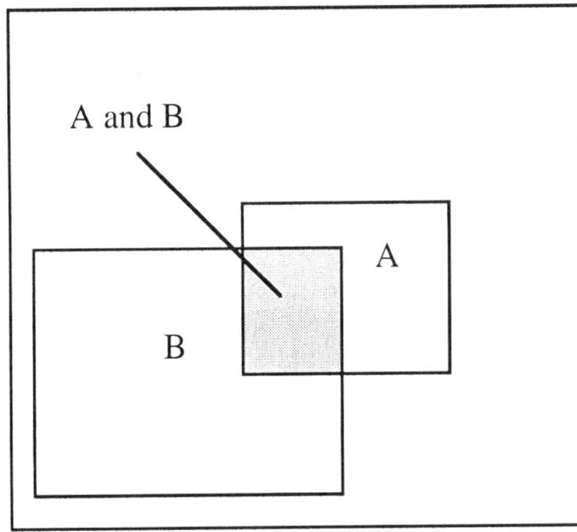

In words, the event A and B is the event that the household selected is both prosperous and educated.

b) The shaded area below is A and B^c

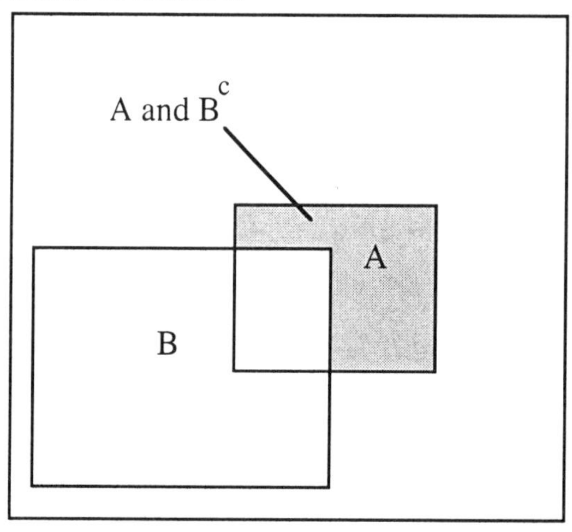

From the diagram, we see that we can write

$$P(\text{A and B}^c) = P(\text{A}) - P(\text{A and B}) = 0.125 - 0.077 = 0.048.$$

In words, the event A and B^c is the event that the household selected is prosperous but not educated.

c) The shaded area below is { A^c and B }

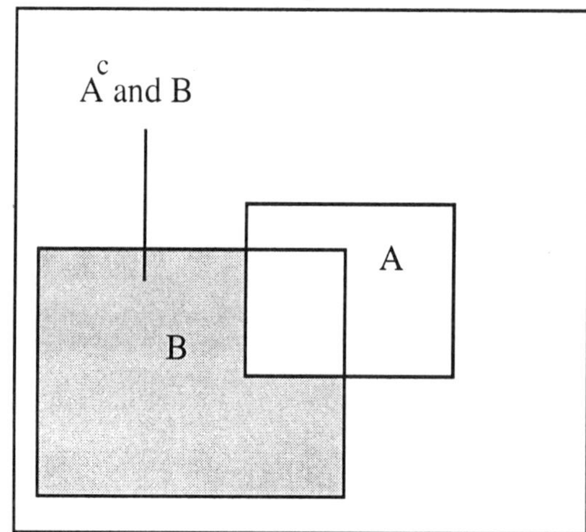

From the diagram, we see that we can write

$$P(\text{A}^c \text{ and B}) = P(\text{B}) - P(\text{A and B}) = 0.237 - 0.077 = 0.160.$$

In words, the event A^c and B is the event that the household selected is not prosperous but is educated.

d) The shaded below area is { A^c and B^c }

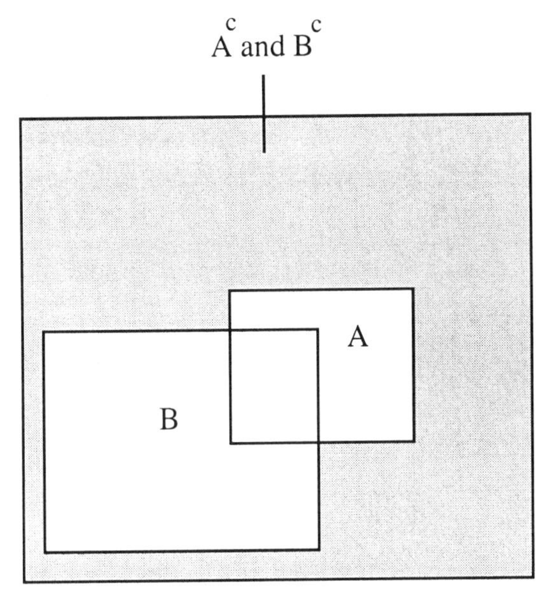

From the diagram, we see that we can write

$$P(A^c \text{ and } B^c) = 1 - P(\text{A or B}) = 1 - 0.285 = 0.715.$$

See exercise 6.37 for the calculation of P(A or B). In words, the event A^c and B^c means the event that the household selected is neither prosperous nor educated.

Exercise 6.41

a) The number of women 65 years and older is found at the bottom of the column labeled "65 and older" and is (in thousands) 18,262. The total number of women in the table is in the lower right corner and is (in thousands) 99,585. The desired probability is thus

(number of women 65 years and older)/(total number of women in table)

= 18262/99585

= 0.1834

b) The desired probability is

(number of married women who are 65 and older)/(number of women 65 years and older)

= 7767/18262

= 0.4253

c) The number of women who are both married and in the 65 and older age group is the entry in the intersection of the row labeled "Married" and the column labeled "65 and older." This number is 7767. The desired probability is

(number of women both "married" and "65 and older") / (total number of women in table)

= 7767/99585

= 0.0780

d) We have found that

P(woman is both "married" and "65 and older") = 0.0780

P(woman is married|woman is 65 and older) = 0.4253

P(woman is 65 and older) = 0.1834

and we see that indeed

P(woman is married|woman is 65 and older)*P*(woman is 65 and older)

= (0.4253)(0.1834)

= 0.0780

= *P*(woman is both "married" and "65 and older")

Exercise 6.53

We know

P(Linda knows the answer) = 0.75

P(Linda is correct | Linda does not know the answer) = 0.20

Our completed tree diagram looks like the following. The numbers in the boxes are those that were given in the problem.

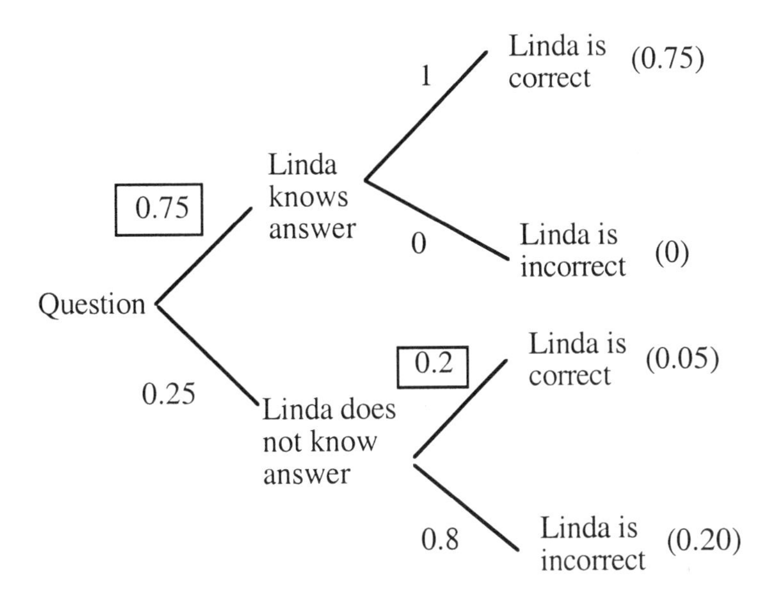

We find that

$$P(\text{Linda is correct}) = (0.75)(1) + (0.25)(0.2) = 0.8$$

Exercise 6.55

For our problem, we take

A = Linda knows answer

B = Linda is correct

Referring to exercise 6.53 and using the multiplication rule, we know that

$$P(\text{A and B}) = P(\text{B}|\text{A})P(\text{A})$$

$$= P(\text{Linda is correct} \mid \text{Linda knows answer})P(\text{Linda knows answer})$$

$$= 1 \times 0.75 = 0.75$$

$$P(\text{A}) = \text{P(Linda is correct)} = 0.8 \text{ (see solution to exercise 6.53)}$$

Thus

$$P(\text{Linda knows answer} \mid \text{Linda is correct}) = \frac{0.75}{0.8} = 0.9375.$$

SELECTED CHAPTER REVIEW EXERCISES

GUIDED SOLUTIONS

Exercise 6.61

KEY CONCEPTS: Independence, events, addition rule, multiplication rule

In many problems, one of the greatest difficulties is deciding on some sort of notation that will allow you to express the problem simply. In this case, the following notation for events should prove helpful. For the husband, you can use the events H_A, H_B, H_O, and H_{AB} to denote the events that the husband has type A, type B, type O, and type AB, respectively. For the wife, the notation for these events is W_A, W_B, W_O, and W_{AB}. With these notational conventions, you can express more complicated events simply and apply the addition and multiplication rules learned in this section to compute the required probabilities.

a) The probability requested is $P(H_B \text{ or } H_O)$. What rule can be used to find this probability?

$P(H_B \text{ or } H_O) =$

b) The probability requested is $P(W_B \text{ and } H_A)$. Remember that blood types of married couples are independent, and the probabilities given for the blood types are valid for both men and women. What rule can be used to find the required probability?

$P(W_B \text{ and } H_A) =$

c) The probability requested is not the same as in part b, although some of the calculations are the same. In this part, we are not told which spouse has type A blood and which spouse has type B blood. The probability requested is given. See whether you can apply the addition and multiplication rules correctly to get to the answer.

$P([W_B \text{ and } H_A] \text{ or } [H_B \text{ and } W_A]) =$

d) First, find the probability that neither person in a randomly chosen couple has type O blood. Let the event $H_{\text{not O}}$ and $W_{\text{not O}}$ correspond to the husband not having type O blood and the wife not having type O blood, respectively. First evaluate

$$P(H_{\text{not O}}) = \qquad\qquad P(W_{\text{not O}}) =$$

Next, use the events $H_{\text{not O}}$ and $W_{\text{not O}}$ to evaluate

$$P(\text{neither has type O}) =$$

You are almost done. How does P(at least one has type O) relate to P(neither has type O)?

COMPLETE SOLUTIONS

Exercise 6.61

a) The probability requested is $P(H_B \text{ or } H_O)$. The events H_B and H_O are disjoint (a person can't have two types of blood), so the addition rule for disjoint events applies:

$$P(H_B \text{ or } H_O) = P(H_B) + P(H_O) = 0.13 + 0.44 = 0.57$$

b) The probability requested is $P(W_B \text{ and } H_A)$. The events W_B and H_A are independent since the blood types of married couples are independent. (Note that these events are not disjoint. It's possible for both of them to occur together. When first learning about probability, students often confuse the ideas of independent and disjoint events.) To evaluate the required probability, we use the multiplication rule for independent events:

$$P(W_B \text{ and } H_A) = P(W_B)P(H_A) = 0.13 \times 0.37 = 0.0481$$

c) The probability requested is $P([W_B \text{ and } H_A] \text{ or } [H_B \text{ and } W_A])$, because, unlike part b, the event does not specify which member of the couple has type A blood and which member has type B blood. The events $[W_B \text{ and } H_A]$ and $[H_B \text{ and } W_A]$ are disjoint, so we can write

$$P([W_B \text{ and } H_A] \text{ or } [H_B \text{ and } W_A]) = P([W_B \text{ and } H_A]) + P([H_B \text{ and } W_A])$$

Each of the probabilities on the right-hand side can be computed as in part b, giving

$$P([W_B \text{ and } H_A]) + P([H_B \text{ and } W_A]) = 0.13 \times 0.37 + 0.13 \times 0.37 = 0.0962$$

d) First, we evaluate $P(H_{\text{not O}}) = 1 - P(H_O) = 1 - 0.44 = 0.56$. Similarly, $P(W_{\text{not O}}) = 0.56$. Now,

$$P(\text{neither has type O}) = P(H_{\text{not O}} \text{ and } W_{\text{not O}}) = P(H_{\text{not O}})P(W_{\text{not O}})$$

where we have used the multiplication rule because $H_{\text{not O}}$ and $W_{\text{not O}}$ are independent. Both the probabilities on the right are equal to 0.56, so we have

$$P(\text{neither has type O}) = P(H_{\text{not O}})P(W_{\text{not O}}) = 0.56 \times 0.56 = 0.3136$$

Finally,

$$P(\text{at least one has type O}) = 1 - P(\text{neither has type O}) = 1 - 0.3136 = 0.6864$$

Although this is the most direct way to do the problem, there are other ways to set it up and arrive at the answer. If you set it up differently and got a different answer, than you probably omitted some cases.

CHAPTER 7

RANDOM VARIABLES

SECTION 7.1

OVERVIEW

A **random variable** is a variable whose value is a numerical outcome of a random phenomenon. The restriction to numerical outcomes makes the description of the probability model simpler and allows us to begin to look at some further properties of probability models in a unified way. If we toss a coin three times and record the sequence of heads and tails, then an example of an outcome would be HTH, which would not correspond directly to a random variable. On the other hand, if we were keeping track of only the number of heads on the three tosses, then the outcome of the experiment would be 0, 1, 2, or 3 and would correspond to the values of the random variable X = number of heads.

The two types of random variables we will encounter are **discrete** and **continuous** random variables. The **probability distribution** of a random variable tells us about the possible values of X and how to assign probabilities to these values. A discrete random variable has a finite number of values, and the probability distribution is a list of the possible values of X and the probabilities assigned to these values. The probability distribution can be given in a table or using a **probability histogram**. For any event described in terms of X, the probability of the event is just the sum of the probabilities of the values of X included in the event.

A continuous random variable takes all values in some interval of numbers. Probabilities of events are determined using a **density curve**. The probability of any any event is the area under the curve corresponding to the values that make up the event. For density curves that involve regular shapes such as rectangles or triangles, we can compute probabilities of events using simple geometrical arguments. The **normal distribution** is another example of a continuous probability distribution, and probabilties of events for normal random variables are computed by standardizing and referring to Table A as was done in Chapter 2.

GUIDED SOLUTIONS

Exercise 7.3

KEY CONCEPTS: Discrete random variables, computing probabilities

a) Write the event in terms of a probability about the random variable *X*. While you can figure out the answer without doing this, it's good practice to start using the notation for random variables.

b) Check to see that all the probabilities are numbers between 0 and 1 and that they sum to 1.

c) Add up the probabilities for values of $X \leq 3$.

d) Add up the probabilities for values of $X < 3$. Make sure that you understand the difference between this answer and what was required for part c.

e) Events about *X* are typically that *X* is bigger than or bigger than or equal to some number, or less than or less than or equal to some number. Write the event in one of these forms.

f) Describe how you would use a TI-83 or Table B (the table of random digits) to represent the probabilities of the various classes. How many digits do you need to use for each trial? What digits would correspond to the event $X \leq 3$? How would you compute the probability of this event?

Exercise 7.5

KEY CONCEPTS: Continuous random variables, computing probabilities

a) As with finding areas under normal curves, it helps to draw a sketch of the density that includes the area corresponding to the probability that you need to evaluate. In this part, you need to find $P(X \leq 49)$ when X has the uniform density curve. The density curve with the area corresponding to this probability is given below. Because it is a rectangular region, the area corresponds to the length $\times$ height $= 0.49 \times 1 = 0.49$, which is the $P(X \leq 0.49)$. Remember for continuous densities, $P(X \leq 0.49) = P(X < 0.49)$.

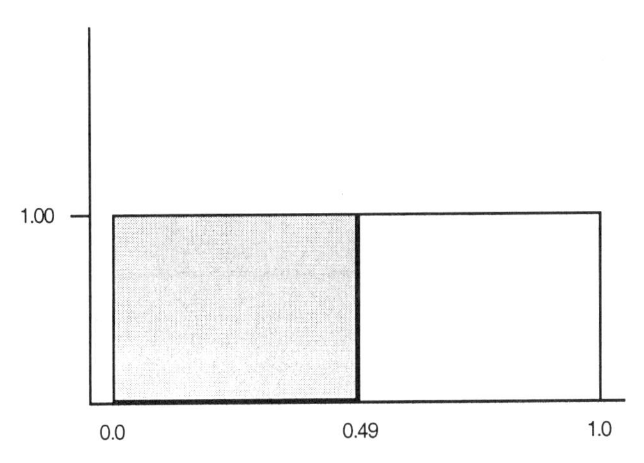

b) Sketch the density and the area you need below.

c) This interval includes values that have no probability because $X \leq 1$. So the values greater than one do not contribute to the probability. Sketch the density and area you need below.

d) Sketch the density and the area you need below.

e) Sketch the density and the area you need below.

f) What is true about the probability of an individual outcome for a continuous random variable?

Exercise 7.9

KEY CONCEPTS: Finding the probability distribution of a random variable

a) The probability of a randomly selected student opposing the funding of interest groups is 0.4 and the probability of favoring it is 0.6. The opinions of different students sampled are independent of each other. So you can use the multiplication rule to find

P(A supports, B supports, and C opposes) =

b) It is easiest to do this by making a table to keep track of the calculations. The first entry is given below. There should be eight lines to the table when you're done. If you've done the calculations correctly, what should be true about the eight probabilities? Don't worry about the column labeled Value of X for now. It will not be needed until part c.

A	B	C	Probability	Value of X
support	support	support	$(0.6)^3 = 0.216$	

c) For each committee listed in the table in part b, find the associated value of X. For the first row, 0 people oppose the funding of interest groups, so the value of X is 0 for a committee with these views. The values of X that can occur are 0, 1, 2, and 3. To find the probability that X takes any of these values, just add up the probabilities of the committees with that value of X. Fill in the table below with your values and make sure that the probabilities sum to 1.

Value of X	Probability

d) If a majority oppose funding, how many people on the committee would have to oppose funding? What does this say about X? Next, use the table you constructed in part c to evaluate this probability.

Exercise 7.15

KEY CONCEPTS: Probabilities for a sample proportion

a) Finding probabilities associated with a sample proportion when we know the mean and standard deviation of the sampling distribution requires first standardizing to a z-score so that we can refer to the table for the standard normal distribution. In this example, the mean is $\mu = 0.15$ (the value of p), and the standard deviation is given as $\sigma = 0.0092$. If you are still uncomfortable doing this type of problem, it is best to continue to draw a picture of a normal curve and the required area as we did in Chapter 2. Otherwise, you can just follow the method of Example 7.4 in this chapter.

$P(\hat{p} \geq 0.16) =$

b) $P(0.14 \leq \hat{p} \leq 0.16) =$

COMPLETE SOLUTIONS

Exercise 7.3

a) $P(X = 5) = 0.01$

b) All the probabilities are between 0 and 1 and the sum of the probabilities is $0.48 + 0.38 + 0.08 + 0.05 + 0.01 = 1.00$

c) For the probability that $X \leq 3$, add the probabilities corresponding to $X = 1$, 2 and 3 to give $P(X \leq 3) = 0.48 + 0.38 + 0.08 = 0.94$.

d) The probability for $X < 3$ includes only values of X that are strictly below 3 that correspond to $X = 1$ and 2. So, $P(X < 3) = 0.48 + 0.38 = 0.86$.

e) The son reaches one of the two higher classes if X is equal to 4 or 5. In terms of a single expression, this event is $X \geq 4$. The $P(X \geq 4) = 0.05 + 0.01 = 0.06$. (Notice that this is the complement of the event in part c, which is why the two probabilities sum to 1.)

f) Because the probabilities are given to two decimal places, we should read the table in sets of two digits. We can represent the various classes as follows.

Son's	Digits in Table B
1	01 – 48
2	49 – 86
3	87 – 94
4	95 – 99
5	00

We would generate two-digit random numbers using a calculator or statistical software, or repeatedly read two-digit numbers in Table B. The event $X \leq 3$ will have occurred if we read any of the numbers 01 to 94. The probability would be the proportion of times we read such numbers.

Exercise 7.5

a) See the guided solution.

b) The shaded ares is $P(X \geq 0.27) = (0.83)(1) = 0.83$.

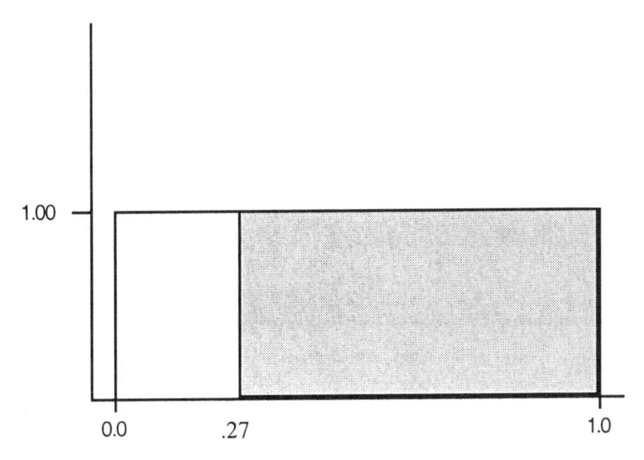

c) $P(0.27 < X < 1.27) = P(0.27 < X < 1.00) = P(X > 0.27)$. This is the probability computed in part b because $P(X > 0.27) = P(X \geq 0.27)$.

d)

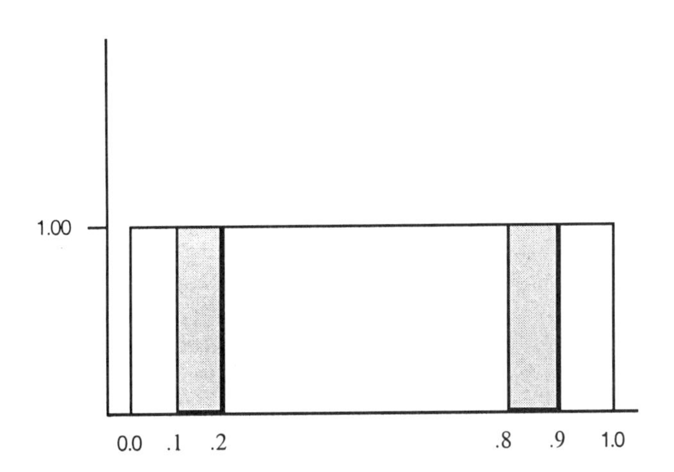

$P(0.1 \leq X \leq 0.2$ or $0.8 \leq X \leq 0.9) = P(0.1 \leq X \leq 0.2) + P(0.8 \leq X \leq 0.9)$, because the two intervals are disjoint.

$P(0.1 \leq X \leq 0.2) = (0.1)(1) = 0.1$ and $P(0.8 \leq X \leq 0.9) = (0.1)(1) = 0.1$.

Putting these together gives $P(0.1 \leq X \leq 0.2$ or $0.8 \leq X \leq 0.9) = 0.1 + 0.1 = 0.2$.

e) In the picture, the area shaded includes all values of X that are not in the interval 0.3 to 0.8.

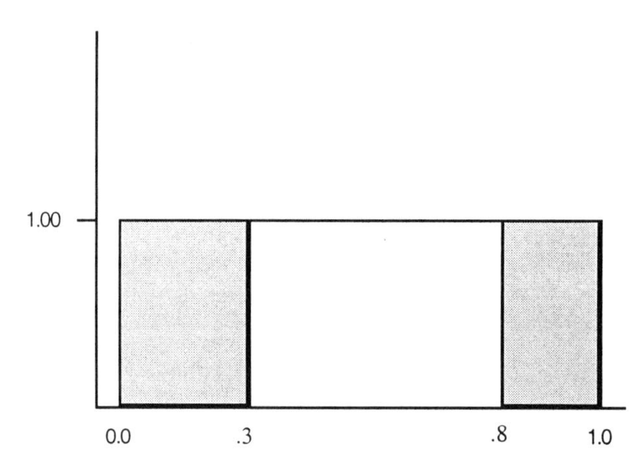

$P(X$ is not in the interval 0.3 to 0.8$) = P(0 \leq X \leq 0.3) + P(0.8 \leq X \leq 1) = 0.3 + 0.2 = 0.5$.

f) Because X has a continuous distribution, the probability of a single point is zero. So, $P(X = 0.5) = 0$.

Exercise 7.9

a) P(A supports, B supports, and C opposes)
= P(A supports) P(B supports) P(C opposes) = (0.6)(0.6)(0.4) = 0.144.

b)

A	B	C	Probability	Value of X
support	support	support	$(0.6)^3 = 0.216$	0
support	support	oppose	$(0.6)^2(0.4) = 0.144$	1
support	oppose	support	$(0.6)^2(0.4) = 0.144$	1
oppose	support	support	$(0.6)^2(0.4) = 0.144$	1
support	oppose	oppose	$(0.6)(0.4)^2 = 0.096$	2
oppose	support	oppose	$(0.6)(0.4)^2 = 0.096$	2
oppose	oppose	support	$(0.6)(0.4)^2 = 0.096$	2
oppose	oppose	oppose	$(0.4)^3 = 0.064$	3
			Total = 1.000	

c) The value of X is given in the table in part b. The possible values of X are 0, 1, 2, and 3. To find the probability that X takes any value, just add up the probabilities of the committees with that value of X. For example, $P(X = 2) = 3(0.096) = 0.288$.

Value of X	Probability
0	0.216
1	0.432
2	0.288
3	0.064

d) A majority says that either 2 or 3 members of the board oppose. If 2 oppose, then $X = 2$, and if 3 oppose, then $X = 3$. So the event is $X \geq 2$, and the required probability is $P(X \geq 2) = 0.288 + 0.064 = 0.352$.

Exercise 7.15

a)

$$P(\hat{p} \geq 0.16) = P\left(\frac{\hat{p} - 0.15}{0.0092} \geq \frac{0.16 - 0.15}{0.0092}\right) = P(Z \geq 1.09) = 1 - 0.8621 = 0.1379$$

b)

$$P(0.14 \leq \hat{p} \leq 0.16) = P\left(\frac{0.14 - 0.15}{0.0092} \leq \frac{\hat{p} - 0.15}{0.0092} \leq \frac{0.16 - 0.15}{0.0092}\right)$$
$$= P(-1.09 \leq Z \leq 1.09) = 0.8621 - 0.1379 = 0.7242$$

SECTION 7.2

OVERVIEW

In Chapter 1, we introduced the concept of the distribution of a set of numbers or data. The distribution describes the different values in the set and the frequency or relative frequency with which those values occur. The mean of the numbers is a measure of the center of the distribution and the standard deviation is a measure of the variability or spread. These concepts also are used to describe features of a random variable X. The probability distribution of a random variable indicates the possible values of the random variable and the probability (relative frequency in repeated observations) with which they occur.

The **mean** μ_X of a random variable X describes the center or balance point of the probability distribution or density curve of X. If X is a discrete random variable having possible values $x_1, x_2, \ldots, x_k$ with corresponding probabilites p_1, $p_2, \ldots, p_k$, the mean μ_X is the average of the possible values weighted by the corresponding probabilities, i.e.,

$$\mu_X = x_1p_1 + x_2p_2 + \ldots + x_kp_k$$

The mean of a continuous random variable is computed from the density curve, but computations require more advanced mathematics. The law of large numbers relates the mean of a set of data to the mean of a random variable and says that the average of the values of X observed in many trials approaches μ_X.

The **variance** σ_X^2 of a random variable X is the average squared deviation of the values of X from their mean. For a discrete random variable

$$\sigma_X^2 = (x_1 - \mu_X)^2p_1 + (x_2 - \mu_X)^2p_2 + \ldots + (x_k - \mu_X)^2p_k.$$

The **standard deviation** σ_X is the positive square root of the variance. The standard deviation measures the variability of the distribution of the random variable X about its mean. The variance of a continuous random variable, like the mean, is computed from the density curve. Again, computations require more advanced mathematics.

The mean and variances of random variables obey the following rules. If a and b are fixed numbers, then

$$\mu_{a + bX} = a + b\mu_X$$

$$\sigma^2_{a + bX} = b^2\sigma^2_X.$$

If X and Y are any two random variables, then

$$\mu_{X+Y} = \mu_X + \mu_Y.$$

If X and Y are independent random variables, then

$$\sigma^2_{X+Y} = \sigma^2_X + \sigma^2_Y.$$

$$\sigma^2_{X-Y} = \sigma^2_X + \sigma^2_Y.$$

GUIDED SOLUTIONS

Exercise 7.17

KEY CONCEPTS: Mean of a random variable

Recall that the average (mean) of X = grade in this course is computed using the formula

$$\mu_X = x_1p_1 + x_2p_2 + \ldots + x_kp_k$$

where the values of x_i and the p_i are given in the following table:

Grade (x)	0	1	2	3	4
Probability (p)	0.10	0.15	0.30	0.30	0.15

Next, use the formula to compute μ_X.

Exercise 7.22

KEY CONCEPTS: Independence, misconceptions about the law of large numbers

The key concept that must be properly understood to answer the questions raised in this problem is the notion of independence. Events A and B are independent if knowledge that A has occurred does not alter our assessment of the probability that B will occur. Do not be misled by misconceptions based on a faulty understanding of the nature of random behavior or a faulty understanding of the law of large numbers (either that runs indicate that a hot streak is in progress and will continue for a while, or that a run of one type of outcome must be immediately balanced by a lack of the outcome for several trials).

Exercise 7.27

KEY CONCEPTS: Variance and standard deviation of a random variable

Recall that the variance of X = grade in this course is computed using the formula

$$\sigma_X^2 = (x_1 - \mu_X)^2 p_1 + (x_2 - \mu_X)^2 p_2 + \ldots + (x_k - \mu_X)^2 p_k$$

where μ_X is the mean of X (calculated in exercise 7.17 as $\mu_X = 2.25$) and the values of x_i and the p_i are given in the following table

Grade (x)	0	1	2	3	4
Probability (p)	0.10	0.15	0.30	0.30	0.15

Next, use the formula to compute $\sigma^2{}_X$. To compute the standard deviation σ_X, take the positive square root of the variance.

Exercise 7.32

KEY CONCEPTS: Rules for means and variances of random variables

a) Follow the same reasoning as in exercise 7.17 and 7.27 to calculate μ_X and σ_X.

b) Calculations are relatively easy using the results of part a if we recall the formulas

$$\mu_{a + bX} = a + b\mu_X$$

$$\sigma^2_{a + bX} = b^2\sigma^2_X.$$

What are a and b in this case?

c) This is just like part b. Again, one must identify a and b.

Exercise 7.39

KEY CONCEPTS: Probability histograms, means, variances, and standard deviations of random variables, comparing probability distributions

The probability histograms consist of bars centered on the number of persons (displayed on the vertical axis and with heights equal to the corresponding probabilities). Use the axes provided below to draw the two histograms.

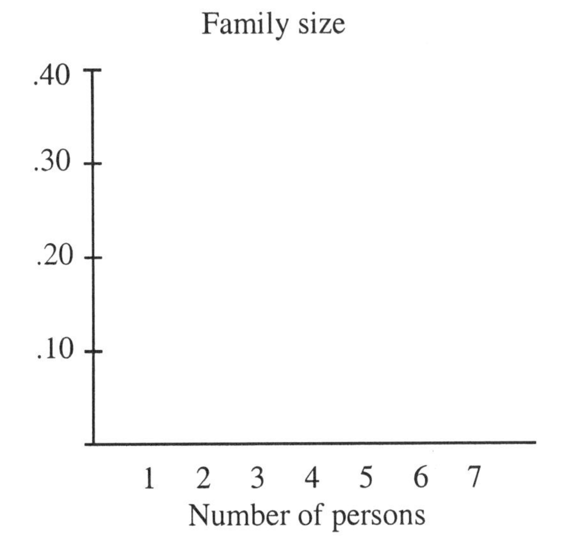

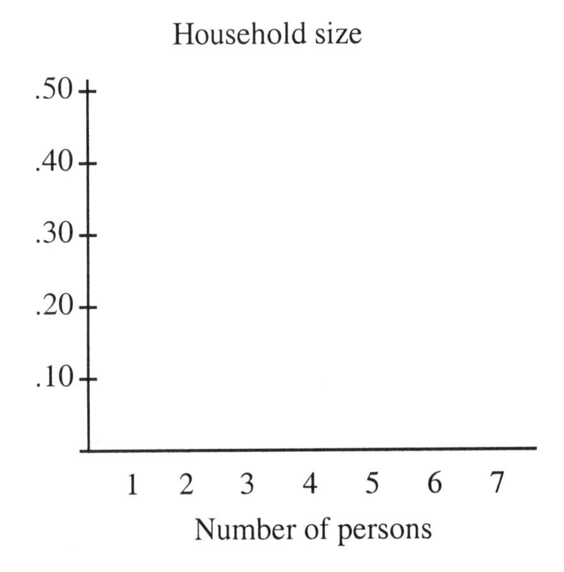

To calculate the mean μ_X and standard deviation σ_X , use the formulas

$$\mu_X = x_1p_1 + x_2p_2 + \ldots + x_kp_k$$

$$\sigma_X^2 = (x_1 - \mu_X)^2p_1 + (x_2 - \mu_X)^2p_2 + \ldots + (x_k - \mu_X)^2p_k.$$

Use your results to compare the two distributions.

COMPLETE SOLUTIONS

Exercise 7.17

We calculate

$$\begin{aligned}\mu_X &= 0(0.10) + 1(0.15) + 2(0.30) + 3(0.30) + 4(0.15)\\ &= 0 + 0.15 + 0.60 + 0.90 + 0.60 = 2.25.\end{aligned}$$

Exercise 7.22

a) Consecutive spins of a fair roulette wheel should be independent. Thus the particular results of previous spins will not change the probability of any particular outcome on the next spin. On the next spin, black is just as likely as red. The gambler's reasoning that red is "hot" fails to recognize that spins are independent. It is based on a false understanding of random behavior.

b) The gambler is wrong again because he is assuming that consecutive cards are independent. This is not the case here. Initially, the deck contains 52 cards, half of which are red and half of which are black. However, each time a card is removed from the deck, the number of of red and black cards remaining changes. For example, if I am dealt five red cards from a deck of 52 cards, the deck now contains only 47 cards, of which 21 are red and 26 are black. The probability that the next card is red is 21/47, which is less than the probability that the next card is black, which is 26/47.

Exercise 7.27

We compute

$$\sigma^2_X = (0 - 2.25)^2(0.10) + (1 - 2.25)^2(0.15) + (2 - 2.25)^2(0.30) + (3 - 2.25)^2(0.30) + (4 - 2.25)^2(0.15)$$

$$= 0.50625 + 0.234375 + 0.01875 + 0.16875 + 0.459375$$

$$= 1.3875$$

and taking the positive square root gives

$$\sigma_X = 1.1779$$

Exercise 7.32

a) We compute

$$\mu_X = 540(0.1) + 545(0.25) + 550(0.3) + 555(0.25) + 560(0.1)$$

$$= 54 + 136.25 + 165 + 138.75 + 56$$

$$= 550$$

$$\sigma^2_X = (540 - 550)^2(0.1) + (545 - 550)^2(0.25) + (550 - 550)^2(0.25) + (555 - 550)^2(0.25) + (560 - 550)^2(0.1)$$

$$= 10 + 6.25 + 0 + 6.25 + 10$$

$$= 32.5$$

$$\sigma_X = \sqrt{32.5} = 5.7$$

b) Here we are interested in $X - 550$ so $a = -550$ and $b = 1$. Thus, we have

$$\mu_{-550 + X} = -550 + 1\mu_X = -550 + 550 = 0$$

$$\sigma^2_{-550 + X} = 1^2\sigma^2_X = 32.5$$

hence

$$\sigma_{-550 + X} = \sqrt{32.5} = 5.7$$

c) Here we want $(9/5)X + 32$ so $a = 32$ and $b = 9/5$. Thus,

$$\mu_{32 + (9/5)X} = 32 + (9/5)\mu_X = 32 + (9/5)550 = 32 + 990 = 1022$$

$$\sigma^2_{32 + (9/5)X} = (9/5)^2\sigma^2_X = (81/25)32.5 = 105.3$$

hence

$$\sigma_{32 + (9/5)X} = \sqrt{105.3} = 10.26.$$

Exercise 7.39

We calculate,

<u>Household size:</u>

$$\begin{aligned}\mu_X &= 1(0.25) + 2(0.32) + 3(0.17) + 4(0.15) + 5(0.07) + 6(0.03) + 7(0.01)\\ &= 0.25 + 0.64 + 0.51 + 0.60 + 0.35 + 0.18 + 0.07\\ &= 2.60\end{aligned}$$

$$\begin{aligned}\sigma^2_X &= (1 - 2.6)^2(0.25) + (2 - 2.6)^2(0.32) + (3 - 2.6)^2(0.17) + \\ &\quad (4 - 2.6)^2(0.15) + (5 - 2.6)^2(0.07) + (6 - 2.6)^2(0.03) + (7 - 2.6)^2(0.01)\\ &= 0.64 + 0.1152 + 0.0272 + 0.294 + 0.4032 + 0.3468 + 0.1936\\ &= 2.02\end{aligned}$$

$$\sigma_X = \sqrt{2.02} = 1.42$$

Family size:

$$\mu_X = 1(0) + 2(.42) + 3(.23) + 4(.21) + 5(.09) + 6(.03) + 7(.02)$$

$$= 0 + 0.84 + 0.69 + 0.84 + 0.45 + 0.18 + 0.14$$

$$= 3.14$$

$$\sigma^2_X = (1 - 3.14)^2(0) + (2 - 3.14)^2(0.42) + (3 - 3.14)^2(0.23) + (4 - 3.14)^2(0.21) + (5 - 3.14)^2(0.09) + (6 - 3.14)^2(0.03) + (7 - 3.14)^2(0.02)$$

$$= 0 + 0.5458 + 0.0045 + 0.1553 + 0.3114 + 0.2454 + 0.2980$$

$$= 1.5604$$

$$\sigma_X = \sqrt{1.5604} = 1.2492$$

Here are the completed probability histograms.

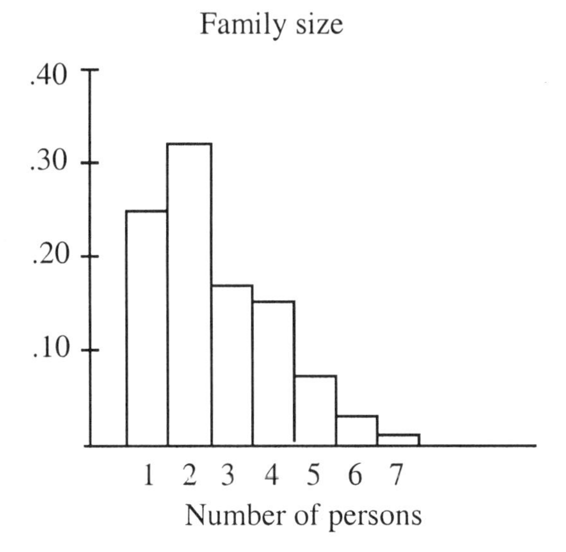

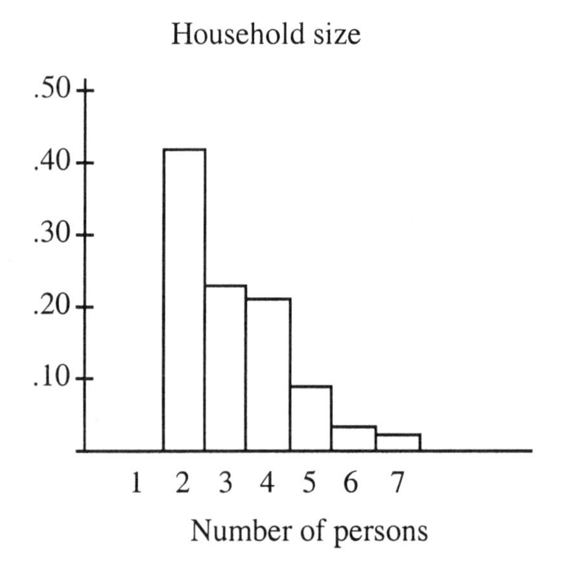

Household size tends to be bigger than family size by about 1. The probability histogram for household size is shifted to the right of that for family size by about 1 unit with little increase in the variability (the standard deviations are fairly close). This is further supported by the fact that the mean household size is larger than the mean family size by a little more than 1.

CHAPTER 8

THE BINOMIAL AND GEOMETRIC DISTRIBUTIONS

SECTION 8.1

OVERVIEW

One of the most common situations giving rise to a count X is the **binomial setting**. The binomial setting consists of four assumptions about how the count was produced. They are

- the number n of observations is fixed;
- the n observations are all independent;
- each observation falls into one of two categories called "success" and "failure";
- the probability of success p is the same for each observation.

When these assumptions are satisfied, the number of successes, X, has a **binomial distribution** with n trials and success probability p. For smaller values of n, the probabilities for X can be found easily using a calculator, statistical software or the exact **binomial probability formula**. The formula is given by

$$P(X = k) = \binom{n}{k} p^k (1-p)^{n-k}$$

where $k = 0, 1, 2, ..., n$, and $\binom{n}{k} = \frac{n!}{k!(n-k)!}$ is called the **binomial coefficient**.

For a binomial random variable X, the **probability density function** (p.d.f.) assigns a probability to each value of X, while the **cumulative distribution function** (c.d.f.) assigns the sum of the probabilities for values less than or equal to X. Most calculators and software will give both the p.d.f. and the c.d.f. for a binomial random variable.

When the population is much larger than the sample, a count X of successes in an SRS of size n has approximately the binomial distribution with n equal to the sample size and p equal to the proportion of successes in the population.

The mean of a binomial random variable X is

$$\mu = np$$

and the standard deviation is

$$\sigma = \sqrt{np(1-p)}\ .$$

GUIDED SOLUTIONS

Exercise 8.1

KEY CONCEPTS: Binomial setting

Four assumptions need to be satisfied to ensure that the count X has a binomial distribution: The number of observations or trials is fixed in advance; each trial results in one of two outcomes; the trials are independent; and the probability of success is the same from trial to trial. Think about how this fits in the binomial setting. What is n? What are the two outcomes, and why might the trials be considered independent?

Exercise 8.2

KEY CONCEPTS: Binomial setting

Four assumptions need to be satisfied to ensure that the count X has a binomial distribution: The number of observations or trials is fixed in advance; each trial results in one of two outcomes; the trials are independent; and the probability of success is the same from trial to trial. In this problem, think about how many observations or trials there are going to be.

Exercise 8.9

KEY CONCEPTS: Binomial probabilities

X is the number of players among the 20 who graduate. According to the university's claim, X should have the binomial distribution with $n = 20$ and $p = 0.8$. You need to find the probability that exactly 10 out of 20 players, graduate, or evaluate $P(X = 10)$. The exact binomial probability formula is given by

$$P(X = k) = \binom{n}{k} p^k (1 - p)^{n-k}$$

where $k = 0, 1, 2, \ldots, n$ and $\binom{n}{k}$ is the binomial coefficient. Plug the appropriate values of n, k and p in the formula to arrive at the answer, or use a calculator or statistical software to evaluate the p.d.f. at $X = 10$.

$P(X = 10) =$

Exercise 8.10

KEY CONCEPTS: Binomial probabilities

a) A count of successes in a SRS of size n has approximately the binomial distribution with n equal to the sample size and p equal to the proportion of successes in the population. What are the values of n and p in this example?

$n =$ ____________________ $p =$ ____________________

b) X is the number of women in the sample of 10 who have never been married. This fits the binomial setting with $n = 10$ trials, and letting "success" correspond to having never been married, we have $p = 0.25$. So X is B (10, 0.25). The probabilities below give the probability density function for X. You should verify these probabilities using either a calculator, statistical software, or the binomial formula. You can use the p.d.f. to evaluate the probabilities for the events described in parts b and c.

k	$P(X = k)$
0	0.056314
1	0.187712
2	0.281568
3	0.250282
4	0.145998
5	0.058399
6	0.016222
7	0.003090
8	0.000386
9	0.000029
10	0.000001

$P(X = 2) =$

c) Write the event "two or fewer have never been married" in terms of X and then use the tabled probabilities to find the desired probability.

Exercise 8.17

KEY CONCEPTS: Mean and standard deviation of a binomial count

You can use the formulas that express the mean and standard deviation of X in terms of n and p. Refer to exercise 8.10 for these values.

mean =

standard deviation =

Exercise 8.21

KEY CONCEPTS: Binomial probabilities, mean and variance of the binomial

a) Let X = number of truthful persons "failing" the lie detector test. X has a binomial distribution with $n = 12$ and $p = 0.2$. So X is $B(12, 0.2)$. The probabilities below give the probability density function for X. You should verify these probabilities using either a calculator, statistical software, or the binomial formula. You can use the p.d.f. given below to evaluate the probabilities for the event described in part a.

k	$P(X = k)$
0	0.068719
1	0.206158
2	0.283468

k	$P(X = k)$ (continued)
3	0.236223
4	0.132876
5	0.053150
6	0.015502
7	0.003322
8	0.000519
9	0.000058
10	0.000004
11	0.000000
12	0.000000

Probability that the lie detector says all 12 are truthful =

Probability that the lie detector says at least one is deceptive =

b) Use the formula that expresses the mean and standard deviation of the binomial in terms of n and p.

mean =
standard deviation =

COMPLETE SOLUTIONS

Exercise 8.1

The number of trials 20 is fixed, each child is a boy or girl, whether or not a child is a boy will not alter the probability that any other children are either girls or boys (excluding identical twins to keep it simple), and the probability of any child being a boy should be the same (about 0.5) for each birth. The binomial distribution should be a good probability model for the number of girls.

Exercise 8.2

Although each birth is a boy or girl, we are not counting the number of successes in a fixed number of births. The number of observations (births) is random. The assumption of a fixed number of observations is violated.

Exercise 8.9

You need to evaluate the formula $P(X = k) = \binom{n}{k} p^k (1-p)^{n-k}$, in which $n = 20$, $k = 10$, and $p = 0.80$.

$$P(X=10)=\binom{20}{10}(0.8)^{10}(0.2)^{10} = \frac{20!}{10!10!}(0.8)^{10}(0.2)^{10} = 0.0020$$

Or you can use a calculator or statistical software to evaluate the p.d.f at $X = 10$ to arrive at this answer.

Exercise 8.10

a) X has approximately a binomial distribution with $n = 10$, the sample size and $p = 0.25$, the proportion of successes (those never having been married) in the population.

b) From the table of the p.d.f., we have $P(X = 2) = 0.281568$. Or you can substitute $n = 10$, $k = 2$, and $p = 0.25$ into the formula giving

$$P(X=2)=\binom{10}{2}(0.25)^2(0.75)^8 = \frac{10!}{2!8!}(0.25)^2(0.75)^8 = (45)(0.0625)(0.10012)$$

$$= 0.2816$$

c) You need to evaluate $P(X \leq 2) = P(X=0)+P(X=1)+P(X=2)$. From the table of the p.d.f., we have

$$P(X \leq 2) = 0.056314 + 0.187712 + 0.281568 = 0.525594,$$

or using the formula, we have

$$P(X \leq 2) = \binom{10}{0}(0.25)^0(0.75)^{10} + \binom{10}{1}(0.25)^1(0.75)^9 + \binom{10}{2}(0.25)^2(0.75)^8$$

$$= 0.0563 + 0.1877 + 0.2816 = 0.5256,$$

in which you need to remember that $0! = 1$ and $(0.25)^0 = 1$ when applying the formulas.

Exercise 8.17

The mean of X is $np = 10(0.25) = 2.5$, and the standard deviation of X is

$$\sqrt{np(1-p)} = \sqrt{10(0.25)(0.75)} = 1.37$$

Exercise 8.21

a) X is B (12, 0.2).

Probability that the lie detector says all 12 are truthful = $P(X = 0) = 0.068719$.

Probability that the lie detector says at least one is deceptive = $P(X \geq 1)$.

You are asked to evaluate $P(X \geq 1)$, the probability that the polygraph says that at least one person telling the truth is deceptive. Using the table of probabilities, you may at first think to add the probabilites for $k = 1, 2, 3, ..., 12$. While this will give the correct answer, in this case it is much simpler to use the rule for complements. $P(X \geq 1) = 1 - P(X = 0) = 1 - 0.068719 = 0.931281$. When determing the complementary event, you must be careful whether an event is of the form greater than or greater than or equal to. The complement of $X \geq 2$ is $X \leq 1$, while the complement of $X > 2$ is $X \leq 2$.

b) The mean is $np = 12(0.2) = 2.4$, and the standard deviation is found using the formula $\sqrt{np(1-p)} = \sqrt{12(0.2)(0.8)} = 1.386$

SECTION 8.2

OVERVIEW

A random variable X has a **geometric distribution** if

- Each observation falls into one of just two categories, which are called "success" and "failure" for convenience.
- The probability p of a success is the same for each observation.
- The observations are independent.
- X counts the number of trials required to obtain the first success.

The first three conditions are the same as for the binomial setting. The difference is that we are counting the number of trials until the first success rather than the number of successes in a given number of trials.

If X has a geometric distribution, then probabilities are calculated using the formula

$$P(X = n) = (1 - p)^{n-1}p.$$

The **mean**, or **expected value**, μ of X is

$$\mu = 1/p$$

and represents the expected number of trials required to get the first success.

The probability that it takes more than n trials to see the first success is

$$P(X > n) = (1 - p)^n$$

GUIDED SOLUTIONS

Exercise 8.24

KEY CONCEPTS: Geometric distributions

The conditions that must be satisfied for a random variable to have a geometric distribution are

- Each observation falls into one of just two categories, which are called "success" and "failure" for convenience.
- The probability p of a success is the same for each observation.
- The observations are independent.
- X counts the number of trials required to obtain the first success.

For each situation described, we need to determine whether any of these conditions are not satisfied. For conditions 2 and 3, you may be able to say only whether the assumptions are reasonable.

a) To get you started, we give a full solution to this part. Notice that

1. The first condition is satisfied because flips fall into only the two categories heads and tails.

2. Getting a tail is a success here. Assuming the coin is not altered between flips and is flipped the same way on each trial, the probability of a success (tail) should be the same for each flip.

3. Again, assuming the coin is not altered between flips and is flipped the same way on each trial, observations (flips) should be independent.

4. We are interested in the number of flips, X, until we observe a tail (success). Thus, X does count the number of trials required to obtain the first success.

Thus, this experiment describes a geometric distribution. The events of interest are tails (success) and heads (failure), a trial is a flip of the coin, and (assuming the coin is fair) the probability of a tail (success) is 1/2.

b) An observation consists of recording whether both of two free throws are made in a one-and-one situation. Answer each of the following.

1. Are there only two outcomes? Think about what we are counting. What constitutes a success?

2. Is it reasonable to assume the probability of success is the same for each observation?

3. Is it reasonable to assume the observations are independent?

4. Are we counting the number of observations until the first success?

If you decided that the experiment describes a geometric distribution, complete the following.

a success is:

a failure is:

a trial is:

the probability of a success is:

If you decided that the experiment does not describe a geometric distribution, what condition is not satisfied?

condition not satisfied is:

c) What constitutes an observation?

To answer the question, complete the following.

1. Are there only two outcomes? Think about what we are counting. What constitutes a success?

2. Is it reasonable to assume the probability of success is the same for each observation?

3. Is it reasonable to assume the observations are independent?

4. Are we counting the number of observations until the first success?

If you decided that the experiment describes a geometric distribution, complete the following.

a success is:

a failure is:

a trial is:

the probability of a success is:

If you decided that the experiment does not describe a geometric distribution, what condition is not satisfied?

condition not satisfied is:

d) What constitutes an observation?

To answer the question, complete the following.

1. Are there only two outcomes? Think about what we are counting. What constitutes a success?

2. Is it reasonable to assume the probability of success is the same for each observation?

3. Is it reasonable to assume the observations are independent?

4. Are we counting the number of observations until the first success?

If you decided that the experiment describes a geometric distribution, complete the following.

a success is:

a failure is:

a trial is:

the probability of a success is:

If you decided that the experiment does not describe a geometric distribution, what condition is not satisfied?

condition not satisfied is:

e) What constitutes an observation?

To answer the question, complete the following.

1. Are there only two outcomes? Think about what we are counting. What constitutes a success?

2. Is it reasonable to assume the probability of success is the same for each observation?

3. Is it reasonable to assume the observations are independent?

4. Are we counting the number of observations until the first success?

If you decided that the experiment describes a geometric distribution, complete the following.

a success is:

a failure is:

a trial is:

the probability of a success is:

If you decided that the experiment does not describe a geometric distribution, what condition is not satisfied?

condition not satisfied is:

Exercise 8.30

KEY CONCEPTS: Geometric distributions, probability distribution table, probability distribution histogram, cumulative distribution histogram

a) To determine whether this is a binomial or geometric setting, ask yourself whether we are interested in the number of red marbles drawn in a fixed number of draws or in the number of draws until the first draw of a red marble. The answer to this question also will help you define X.

b) Use your calculator to compute these probabilities. On the TI-83, to compute the probability of drawing a red marble on the second draw you will need to use the geometric p.d.f. The probability of drawing a red marble is 20/35 (Can you see why?) and the appropriate key sequence is

2nd/DISTR/ALPHA/D/geometpdf(20/35,2) followed by ENTER

What do you get for the probability?

To compute the probability of drawing a red marble by the second draw (i.e., either on the first or second draw), the appropriate key sequence is

2nd/DISTR/ALPHA/E/geometcdf(20/35,2) followed by ENTER

What do you get for the probability?

The probability that it would take more than two draws to get a red marble can be computed from one of the above answers and the rules of probability.

What do you get for the probability?

c) Refer to page 440 in the textbook for information on the appropriate commands.

Place the results of these commands in the following columns.

$X(L_1)$	p.d.f.(L_2)	c.d.f.(L_3)

d) Use 2nd/STAT PLOT to make the desired plots. Be sure to set up your window appropriately (see page 440 of your textbook).

COMPLETE SOLUTIONS

Exercise 8.24

a) A complete solution was given in the guided solutions.

b) An observation consists of recording whether both of two free throws are made in a one-and-one situation. Note that

1. A success is making both free throws. A failure is any other outcome. Thus, there are only two outcomes.

2. This condition is not easy to assess. It may not be unreasonable to assume the probability of making both free throws is the same for each observation. However, one could also argue that, due to pressure or physical exhaustion, a player's ability to make a free throw changes over the course of a game and is not constant.

3. This condition is also difficult to assess. It may not be unreasonable to treat observations as independent. However, some people believe in "hot hands" and would argue that your confidence increases as you make more shots and, as a result, the probability of making both free throws changes depending on how you have performed previously.

4. We are simply counting the number of times a player makes both free throws in a one-and-one foul-shooting situation. We are *not* counting the number of observations until a player makes both free throws for the first time.

This experiment does not describe a geometric distribution, and the

condition not satisfied is: condition 4 is clearly violated. Conditions 2 and 3 may not be satisfied either.

c) An observation consists of drawing a card and noting whether or not it is a jack.

1. The outcomes are "the card is a jack" (success) and "the card is not a jack" (failure), so there are two outcomes.

2. Assuming cards are drawn at random, the probability of success should be the same for each observation, because the card is replaced and so we are always drawing from a complete deck.

3. If cards are drawn at random, observations should be independent because the card is replaced after each draw.

4. We are counting the number of observations (draws) until the first jack (success) is observed.

The experiment describes a geometric distribution, and

a success is: drawing a jack

a failure is: drawing any other card other than a jack

a trial is: drawing a card, observing whether it is a jack, and replacing the card in the deck.

the probability of a success is: 4/52 = 1/13 because there are 52 cards in the deck, four of which are jacks.

d) An observation consists of buying a ticket and determining whether it wins (all six numbers match those drawn by the lottery representative).

1. There are two outcomes. These are "the ticket is a winner" (success) and "the ticket is not a winner" (failure).

2. If the lottery is conducted fairly (numbers are selected randomly), all tickets should have the same chance of winning every time the lottery is played, and so all observations will have the same probability.

3. If the lottery is fair (numbers selected randomly), selection of winning numbers should be independent.

4. We are interested in how long it takes until the first winning ticket is purchased.

The experiment describes a geometric distribution, and

a success is: purchasing a winning ticket

a failure is: purchasing a non-winning ticket.

a trial is: purchasing a ticket and determining whether it is a winner.

the probability of a success is: $1 \Big/ \binom{44}{6}$. This is because 6 numbers out of 44 are selected (without replacement). There are $\binom{44}{6}$ ways of choosing these 6 numbers, all of which are equally likely.

e) An observation consists of drawing a marble and observing whether it is red.

1. There are two outcomes. These are "the marble is red" (success) or the marble is blue (failure).

2. Marbles are *not* replaced, so the probability of a success on any given draw depends on which marbles have already been drawn. However, *without* any knowledge of which marbles have already been drawn, we would compute the probability of a success on any particular draw to be 10/15 = 2/3. So it is true that the probability of success is the same for

each draw in the absence of information of the outcomes of previous draws.

3. Marbles are *not* replaced, so draws are not independent. The probability of the next draw depends on what marbles have been drawn.

4. We are not counting the number of draws until the first red marble. Instead we are counting the number of draws until the third red marble is drawn.

The experiment does not describe a geometric distribution, and the

condition not satisfied is: conditions 3 and 4 are not satisfied.

Exercise 8.30

a) This is a geometric setting because we are interested in how many draws are needed until a red marble is drawn. The random variable X would be

X = the number of draws until the first red marble is drawn.

b) Using the TI-83 calculator key sequences described in the Guided Solution, we find that

Probability of drawing red marble on second draw = $P(X = 2) = 0.2448979592$

Probability of drawing red marble by second draw = $P(X \leq 2) = 0.8163265306$.

The remaining probability is computed as follows.

Probability that it would take more than two draws to get a red marble

$$= P(X > 2) = 1 - P(X \leq 2) = 1 - 0.8163265306 = 0.1836734694.$$

c) To install the first 20 values of X into list L_1, use the command

seq(X,X,1,20,1)→L_1

Recall that seq is obtained by the key sequence 2nd/LIST/OPS/5.

To install the corresponding probabilities into list L_2, use the command

geometpdf(20/35,L_1)→L_2.

To install the cumulative probabilities into list L_3, use the command

geometcdf(20/35,L_1)→L_3.

We copy the results in the table below (rounded to three decimal places) placing the values of X, the p.d.f., and the c.d.f. in columns (rather than rows)

X(L_1)	p.d.f.(L_2)	c.d.f.(L_3)
1	0.5714	0.5714
2	0.2449	0.8163
3	0.1050	0.9213
4	0.0450	0.9663
5	0.0192	0.9855
6	0.0083	0.9938
7	0.0035	0.9973
8	0.0015	0.9989
9	0.0007	0.9995
10	0.0003	0.9998
11	0.0001	0.9999
12	0	1
13	0	1
14	0	1
15	0	1
16	0	1
17	0	1
18	0	1
19	0	1
20	0	1

d) Here are the plots.

p.d.f. histogram

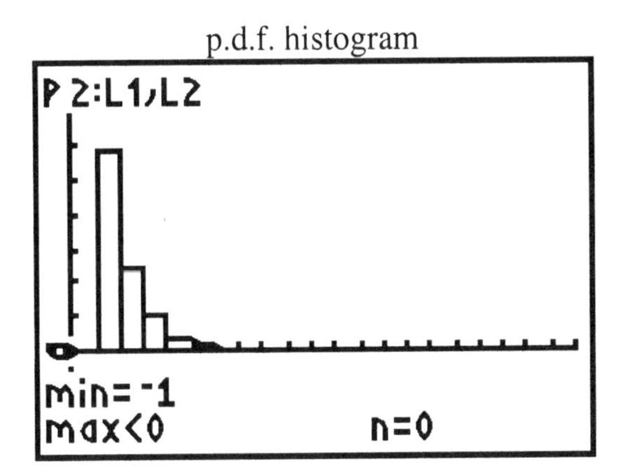

c.d.f. histogram

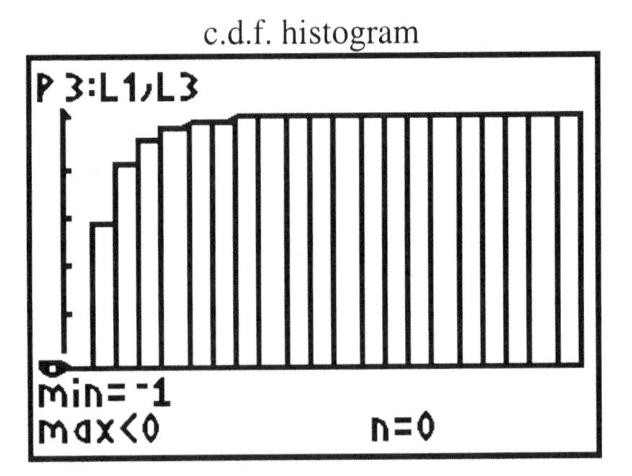

CHAPTER 9

SAMPLING DISTRIBUTIONS

SECTION 9.1

OVERVIEW

Statistical inference is the technique that allows us to use the information in a sample to draw conclusions about the population. To understand the idea of statistical inference, it is important to understand the distinction between **parameters** and **statistics.** A **statistic** is a number we calculate based on a sample from the population—its value can be computed once we have taken the sample, but its value varies from sample to sample. A statistic is generally used to estimate a population **parameter**, which is a fixed but unknown number that describes the population.

The variation in a statistic from sample to sample is called **sampling variability**. It can be described through the **sampling distribution** of the statistic, which is the distribution of values taken by the statistic in all possible samples of the same size from the population. The sampling distribution can be described in the same way as the distributions we encountered in Chapter 1. Three important features of a sampling distribution are

- a measure of center
- a measure of spread
- a description of the shape of the distribution

The properties and usefulness of a statistic can be determined by considering its sampling distribution. If the sampling distribution of a statistic is centered (has its mean) at the value of the population parameter, then the statistic is **unbiased** for this parameter. This means that the statistic tends to neither overestimate nor underestimate the parameter.

Another important feature of the sampling distribution is its spread. If the statistic is unbiased and the sampling distribution has little spread or variability, then the statistic will tend to be close to the parameter it is estimating for most samples. The variability of a statistic is related to both the sampling design and the sample size n. Larger sample sizes give smaller spread (better estimates) for any sampling design. An important feature of the spread is that as long as the population is much larger than the sample (at least ten times), the spread of the sampling distribution will depend primarily on the sample size, not the population size.

GUIDED SOLUTIONS

Exercise 9.3

KEY CONCEPTS: Statistics and parameters

In deciding whether a number represents a parameter or a statistic, you need to think about whether it is a numerical characteristic of the population of interest or whether it is a numerical characteristic of the particular sample that was selected. Statistics vary from sample to sample; parameters are fixed numerical characteristics of the population.

Exercise 9.9

KEY CONCEPTS: Bias and variability of a sampling distribution

If a statistic is unbiased, then the mean of its sampling distribution should be equal to the population mean. In terms of the histograms given, if the statistic displays little bias, the center of the histogram (balance point) should be close to the population parameter indicated by the arrow. Which sampling distributions display little bias? The variability of the sampling distribution refers to its spread. Which of the sampling distributions show a great deal of spread (variability) and which do not?

Exercise 9.12

KEY CONCEPTS: Sampling distributions

a) The table of random numbers contains the 10 digits, 0, 1, 2, ..., 9, which are "equally likely" to occur in any position selected at random from the table. If we want an egg mass to be present 20% of the time, then two digits correspond to the presence of an egg mass and the remaining eight digits correspond to the absence of an egg mass. Does it matter which two digits correspond to the presence of an egg mass?

While any two digits could be used, so that everyone does the same thing, let the occurence of the digits 0 or 1 correspond to the presence of an egg mass, and the remaining digits correspond to the absence. Also, let's start on line 128 of the table

$\underline{1}5689 \quad \underline{1}4227$

These 10 random digits correspond to our 10 sample areas. There are two sample areas with egg masses (corresponding to a digit of 0 or 1), so that $\hat{p}$ = 0.2 for this sample.

b) For this part of the problem, everyone will be taking their 20 samples from different parts of the random number table. Many of you may know how to get random samples from your calculators or computer software. Work with your own samples here. Your answers will not agree exactly with that given in the complete solution, although the general pattern should be similar. If everyone took 2000 samples instead of 20, would the sampling distributions from person to person show more or less agreement? Fill in the values of $\hat{p}$ from your 20 samples in the table below.

sample	$\hat{p}$
1	
2	
3	
4	
5	
6	
7	
8	
9	
10	

sample	$\hat{p}$
11	
12	
13	
14	
15	
16	
17	
18	
19	
20	

Next, draw the stemplot below.

c) How does the mean of the sampling distribution of $\hat{p}$ relate to the population parameter p?

d) What has changed in this setting? How does this affect the mean of the sampling distribution?

COMPLETE SOLUTIONS

Exercise 9.3

43 unlisted numbers is a property of the sample of 100 numbers dialed. It is the value of a statistic. **52%** of all Los Angeles phones unlisted is a property of the population of phones and is the value of a parameter.

Exercise 9.9

a) The center of this histogram is clearly far to the left of the population parameter, indicating large bias. In addition, relative to the four sampling distributions given, this one is quite spread out indicating large variability in the sampling distribution.

b) The center of the histogram appears to be slightly to the right of the population parameter, indicating little bias. In addition, there is little variability in this sampling distribution.

c) The center of the histogram and the population parameter are quite close indicating little bias. The histogram is fairly spread out showing large variability in the sampling distribution.

d) The center of the histogram is nowhere near the population parameter, indicating large bias. However, the histogram shows little variability in the sampling distribution.

Exercise 9.12

a) Done in the guided solution

b) These are the values of $\hat{p}$ in the 20 samples we obtained using the computer to generate random digits, and the stemplot for the 20 values of $\hat{p}$.

sample	$\hat{p}$
1	0.1
2	0.1
3	0.0
4	0.1
5	0.0
6	0.4
7	0.0
8	0.3
9	0.2
10	0.0
11	0.2
12	0.0
13	0.4
14	0.2
15	0.4
16	0.2
17	0.2
18	0.1
19	0.3
20	0.2

```
0.0 | 00000
0.1 | 0000
0.2 | 000000
0.3 | 00
0.4 | 000
```

The mean of the 20 values of $\hat{p}$ is 0.17. Examining the stemplot, the shape looks fairly symmetric with a center near 0.2. Your stem and leaf plot may look quite different from this—with only 20 samples, the simulated sampling distributions might vary quite a bit from person to person. If everyone took 2000 samples, which would require the sampling be done using a computer, then the shapes of the distributions would be quite similar from person to person.

c) The mean of the sampling distribution of $\hat{p}$ is $p = 0.2$. In our simulation, the mean of the 20 samples is 0.17 and agrees fairly well with the theoretical mean of the sampling distribution.

d) Because the mean of the sampling distribution of $\hat{p}$ is p, the mean of the sampling distribution now would be 0.4.

SECTION 9.2

OVERVIEW

When we want to learn about a **population proportion** p, we can take an SRS from the population and use the **sample proportion**

$$\hat{p} = \frac{\text{count of successes in the sample}}{\text{count of observations in the sample}}$$

to estimate p. The sampling distribution of $\hat{p}$ describes how the value of $\hat{p}$ varies in all possible samples from the population.

Here are the basic facts about the sampling distribution of $\hat{p}$ from an SRS of size n taken from a large population.

- The sampling distribution of $\hat{p}$ is approximately normal when the sample size n is large.
- The **mean** of the sampling distribution is exactly p.
- The **standard deviation** of the sampling distribution is $\sqrt{p(1-p)/n}$.

We can apply the formula for the standard deviation of the sampling distribution provided the population is at least 10 times as large as the sample. The normal distribution should provide a good approximation to the sampling distribution of $\hat{p}$ when the values of n and p satisfy $np > 10$ and $n(1\text{-}p) > 10$.

GUIDED SOLUTIONS

Exercise 9.17

KEY CONCEPTS: Normal approximation for proportions, effect of sample size on accuracy

If $\hat{p}$ is to be within 0.02 of $p = 0.15$, then $\hat{p}$ must be between 0.13 and 0.17. So the probability of $\hat{p}$ being within $\pm 2\%$ of 0.15 is $P(0.13 \le \hat{p} \le 0.17)$. When n = 200, approximate this probability using the normal approximation to the sampling distribution of a proportion. Find the mean and standard deviation of the sampling distribution, and then standardize to use Table A. A picture may help.

mean of the sampling distribution of $\hat{p} = p =$

standard deviation of the sampling distribution of $\hat{p} = \sqrt{p(1-p)/n} =$

Computing $P(0.13 \leq \hat{p} \leq 0.17)$ is a normal probability calculation like those in Chapter 2. You may want to review the material there in order to refresh your memory on how to do such calculations, as they form the basis of solving many of the remaining exercises in this chapter. This can be done using your calculator (see exercise 2.34) or by hand. To compute the probability by hand, we standardize the numbers 0.13 and 0.17 (compute their z-scores) by subtracting the mean of the sampling distribution of $\hat{p}$ and dividing the result by standard deviation of the sampling distribution of $\hat{p}$. Next, use Table A to determine the area between the two z-scores. It may be helpful to draw a normal curve and the desired area to help you visualize the area.

$P(0.13 \leq \hat{p} \leq 0.17) =$

Repeat this for $n = 800$ and $n = 3200$. What general conclusion can you draw about the effect of sample size on your calculations?

Exercise 9.20

Key Concepts: Normal approximation for proportions

a) This question does not require computing probabilities. From the information given, what is the value of $\hat{p}$ for these data?

b) We are told that $p = 0.9$ in the population, and an SRS of 100 from the 5000 orders is taken. Because our population (5000) is more than 10 times the size of the sample (100), the formula $\sqrt{p(1-p)/n}$ can be used for the standard deviation of $\hat{p}$. You need to compute a probability about $\hat{p}$, namely $P(\hat{p} \leq 0.86)$. Because $np = 100(0.9) = 90$ and $n(1 - p) = 100(0.1) = 10$ are both greater than or equal to 10, we can use the normal approximation. (However, because n

and p just meet the criteria, you might expect the normal approximation not to be that accurate.) To use the normal approximation, the mean and standard deviation of $\hat{p}$ must first be computed. It may also be helpful to draw a normal curve and the desired area to help you solve the problem.

mean of the sampling distribution of $\hat{p}$ =
standard deviation of the sampling distribution of $\hat{p}$ =

$P(\hat{p} \leq 0.86) =$

c) What does sampling variability tell us?

COMPLETE SOLUTIONS

Exercise 9.17

When $n = 200$, the mean is $p = 0.15$ and the standard deviation is

$$\sqrt{\frac{p(1-p)}{n}} = \sqrt{\frac{0.15(1-0.15)}{200}} = 0.0252.$$

$$P(0.13 \leq \hat{p} \leq 0.17) =$$

$$P\left(\frac{0.13-0.15}{0.0252} \leq \frac{\hat{p}-0.15}{0.0252} \leq \frac{0.17-0.15}{0.0252}\right) = P(-0.79 \leq Z \leq 0.79)$$

$$= 0.7852 - 0.2177 = 0.5681.$$

You may obtain a similar, but not identical, answer on your calculator. Differences are due to roundoff in the above calculations.

When $n = 800$, the mean is $p = 0.15$ and the standard deviation is

$$\sqrt{\frac{p(1-p)}{n}} = \sqrt{\frac{0.15(1-0.15)}{800}} = 0.0126$$

$$P(0.13 \leq \hat{p} \leq 0.17) =$$

$$P\left(\frac{0.13-0.15}{0.0126} \leq \frac{\hat{p}-0.15}{0.0126} \leq \frac{0.17-0.15}{0.0126}\right) = P(-1.59 \leq Z \leq 1.59)$$

$$= 0.9441 - 0.0559 = 0.8882.$$

Again, you should obtain a similar (but not identical) answer if you use a calculator.

When $n = 3200$, the mean is $p = 0.15$ and the standard deviation is

$$\sqrt{\frac{p(1-p)}{n}} = \sqrt{\frac{0.15(1-0.15)}{3200}} = 0.0063.$$

$$P(0.13 \le \hat{p} \le 0.17) =$$

$$P\left(\frac{0.13-0.15}{0.0063} \le \frac{\hat{p}-0.15}{0.0063} \le \frac{0.17-0.15}{0.0063}\right) = P(-3.17 \le Z \le 3.17)$$

$$= 0.9992 - 0.0008 = 0.9984.$$

The general conclusion is that for larger samples, the sample proportion is more likely to be close to the true proportion or equivalently that larger samples give more accurate results due to smaller sampling variability.

Exercise 9.20

a) $\hat{p} = 86 / 100 = 0.86$.

b) For $\hat{p}$, the mean is $p = 0.9$ and the standard deviation is

$$\sqrt{\frac{p(1-p)}{n}} = \sqrt{\frac{0.9(1-0.9)}{100}} = 0.03$$

$P(\hat{p} \le 0.86) = P\left(\frac{\hat{p}-0.9}{0.03} \le \frac{0.86-0.9}{0.03}\right) = P(Z \le -1.33) = 0.0918$, where you need to remember that this is an approximation to $P(\hat{p} \le 0.86)$ The exact answer can be computed using special software and gives $P(\hat{p} \le 0.86) = 0.1239$. Because n and p just meet the criteria to use the approximation, we might not expect the answer using the approximation to be very close to the exact answer in this case.

c) If the true percentage is 90%, the percentage $\hat{p}$ in the sample won't be exactly equal to $p = 0.9$ every time. It will fluctuate around this value according to its sampling distribution. It was shown that a value as small or smaller than 0.86 would occur approximately 10% of the time, which is not that unusual. Our data do not contradict the claim, but instead illustrate natural sampling variability.

SECTION 9.3

OVERVIEW

To learn about the **population mean** μ for some variable, we can use the **sample mean** $\bar{x}$ as an estimate when the observations are an SRS from a large population. As with all sampling distributions, the sampling distribution of $\bar{x}$ describes how the value of $\bar{x}$ varies in all possible samples from the population.

Here are the basic facts about the sampling distribution of $\bar{x}$ from an SRS of size n taken from a population where the mean is μ and the standard deviation is σ:

- $\bar{x}$ is an **unbiased** estimate of the population mean μ, so the **mean** of the sampling distribution of $\bar{x}$ is the population mean μ.

- The **standard deviation** is $\sigma/\sqrt{n}$ where σ is the population standard deviation. We can apply the formula for the standard deviation of the sampling distribution provided the population is at least 10 times as large as the sample. The formula shows that there is less variation in averages than in individuals and that averages based on larger samples are less variable than those based on smaller samples.

- The shape of the distribution of the sample mean depends on the shape of the population distribution. If the population were normal, $N(\mu, \sigma)$, then the sample mean has normal distribution, $N(\mu, \sigma/\sqrt{n})$. Also, for large samples the **Central Limit Theorem** tells us the sample mean has *approximately* a normal distribution, $N(\mu, \sigma/\sqrt{n})$.

The **law of large numbers** also tells us about the behavior of $\bar{x}$ as the sample size increases. The law of large numbers describes how the mean of many observations will get closer and closer to the mean of the population.

GUIDED SOLUTIONS

Exercise 9.27

KEY CONCEPTS: Mean and standard deviation of a sample mean

The mean and standard deviation of the sample mean can be expressed in terms of the population mean and standard deviation and the sample size. First, identify these three quantities in this problem.

$\mu =$

$\sigma =$

$n =$

Mean of the average score =

Standard deviation of the average score =

Do these results depend on the individual scores having a normal distribution?

Exercise 9.31

KEY CONCEPTS: The sampling distribution of the sample mean, normal probability calculations

a) You may wish to review normal probability calculations in Chapter 2 to refresh your memory. What is the z-score of 295?

z-score =

Shade in the desired area in the figure below.

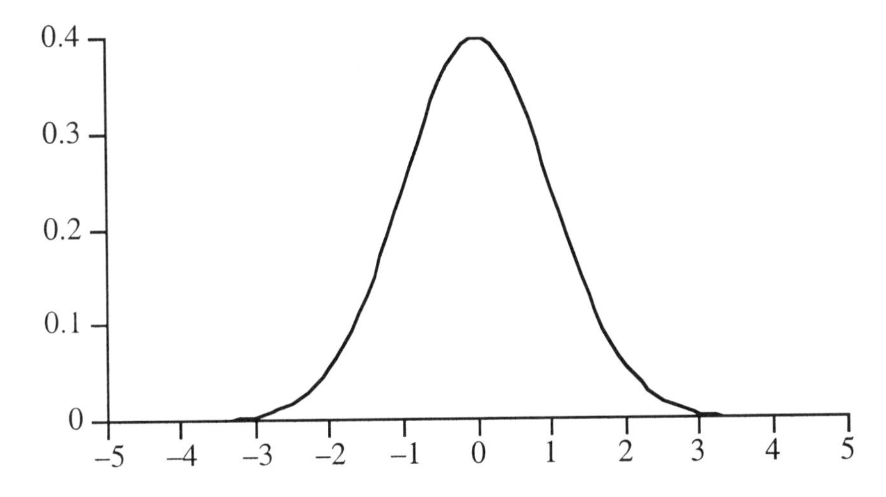

Next, use Table A or your calculator to find this area.

area =

b) Next, we are interested in probabilities concerning the mean of the contents of six bottles. What do we know about the sampling distribution of the mean based on samples of size 6 from a *normal* population? What, therefore, is the appropriate sampling distribution for the probability requested?

sampling distribution =

Based on this sampling distribution, what is the z-score of 295 here?

z-score =

Draw a picture of the desired area under a standard normal curve and use Table A or your calculator to find this area.

EXERCISE 9.32

KEY CONCEPTS: The sampling distribution of the sample mean, normal probability calculations.

a) Recall that the sampling distribution of a sample mean of observations from a normal population with mean μ and standard deviation σ is normal with

mean of the sampling distribution of $\bar{x}$ =

standard deviation of the sampling distribution of $\bar{x}$ =

The sampling distribution is $N($, $)$.

b) This is just a normal probability calculation, like those we did in Chapter 2. We want the probability that the sample mean $\bar{x}$ is 124 mg or higher. To compute this probability, we must first standardize the number 124 (compute its z-score) by subtracting the mean of the sampling distribution of $\bar{x}$ and dividing the result by the standard deviation of the sampling distribution of $\bar{x}$, which were computed in part a. Next, use Table A or your calculator to determine the area to the right of this z-score under a standard normal curve. You may wish to draw a picture of the standard normal to help you visualize the desired area.

Exercise 9.40

KEY CONCEPTS: The sampling distribution of the sample mean, backward normal probability calculations

We are told that the distribution of individual scores at Southwark Elementary School is approximately normal with mean $\mu = 13.6$ and standard deviation $\sigma = 3.1$. To find L, Mr. Lavin needs to first determine the sampling distribution of the mean score $\bar{x}$ of $n = 22$ children. What is this sampling distribution?

To complete the problem, you need to find L such that the probability of $\bar{x}$ being below L is only 0.05. We did this type of problem in Chapter 2. First, refer to Table A to find the value z such that the area to the left of z under a standard normal curve is 0.05. What is this value?

This value z is the z-score of L. This means $z = (L - \text{mean})/\text{standard deviation}$, in which the mean and standard deviation referred to are those of the sampling distribution of $\bar{x}$. Solve this equation for L.

You also can do these backward calculations using a calculator. See exercise 2.35 for details.

COMPLETE SOLUTIONS

Exercise 9.27

From the information in the problem, we have

$\mu = 18.6$
$\sigma = 5.9$
$n = 76$

Mean of the average score = $\mu = 18.6$.

Standard deviation of the average score = $\sigma/\sqrt{n} = 5.9/\sqrt{76} = 0.677$.

These results are true no matter what the population distribution is, so the fact that the individual scores have a normal distribution is not required to obtain the mean and the standard deviation of the average score.

Exercise 9.31

a) The probability of interest is displayed as the shaded are in the figure below.

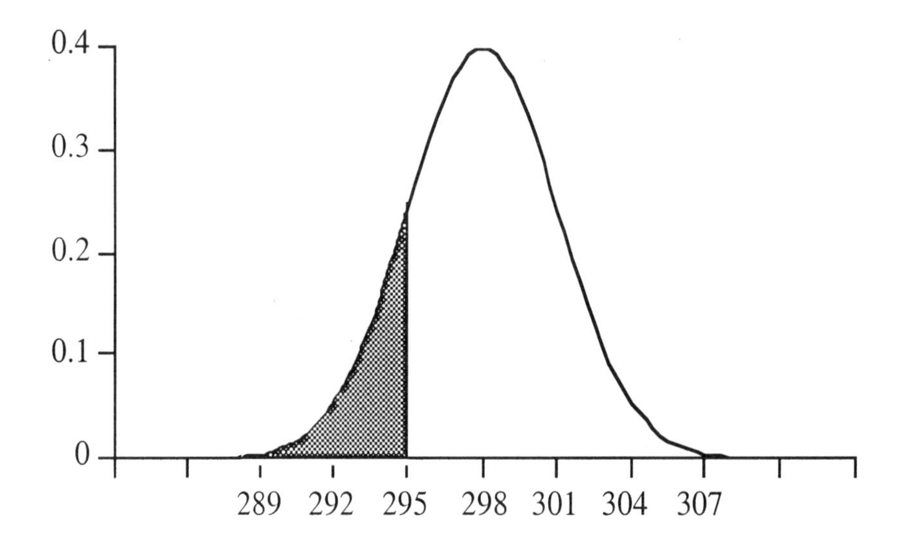

To find this area, we must compute the z-score of 295. Because the contents follow a normal distribution with mean $\mu = 298$ and standard deviation $\sigma = 3$, the z-score of 295 is

$$\begin{aligned} z\text{-score} &= \frac{x-\mu}{\sigma} \\ &= \frac{295-298}{3} \\ &= \frac{-3}{3} = -1. \end{aligned}$$

In terms of areas under the standard normal, the area of interest is the area to the left of –1, which can be read directly from Table A. From Table A, we find that the area to the left of –1 under a standard normal curve is 0.1587. Therefore, the probability that an individual bottle contains less than 295 ml is 0.1587.

b) For questions about probabilities associated with the mean contents of the bottles in a six-pack, we need to determine the sampling distribution of the mean of six bottles. The contents of individual bottles vary according to a normal distribution with mean $\mu = 298$ and standard deviation $\sigma = 3$. Because we have a random sample from a normal distribution, this sampling distribution for the mean of the contents of six bottles is also normal with the mean $\mu = 298$ (the same as for individual bottles) and standard deviation

$$\frac{\sigma}{\sqrt{n}} = \frac{3}{\sqrt{6}} = 1.22.$$

Next, to compute the probability that the mean contents of the bottles in a six-pack is less than 295, again we need to find the z-score of 295. This time, we find the z-score is

$$z\text{-score} = \frac{295 - 298}{1.22}$$
$$= \frac{-3}{1.22} = -2.46.$$

In terms of areas under the standard normal, the area of interest is now the area to the left of –2.46. This area can be read directly from Table A. From Table A, we find that the area to the left of –2.46 under a standard normal curve is 0.0069. Therefore, the probability that the mean contents of the bottles in a six-pack is less than 295 ml is 0.0069.

EXERCISE 9.32

a) We are given that the population has mean $\mu = 123$ mg, has standard deviation $\sigma = 0.08$ mg, and that the population is normal. Thus, we know that the sampling distribution of this mean also will be normal. In addition, the sampling distribution will have

mean of the sampling distribution of $\bar{x} = \mu = 123$ mg

standard deviation of the sampling distribution of $\bar{x} = \frac{\sigma}{\sqrt{n}} = \frac{0.08}{\sqrt{3}} = 0.046$

The sampling distribution of $\bar{x}$ is N(123 mg, 0.046 mg).

b) Using the mean and standard deviation of the sampling distribution of $\bar{x}$ found in part a, the z-score of 124 is

$$z\text{-score} = \frac{124 - 123}{0.046} = 21.74.$$

The area to the right of this under a standard normal curve is not in Table A because the z-score exceeds the values given in the Table. It is essentially 0.

EXERCISE 9.40

The sampling distribution of the mean score $\bar{x}$ of 22 children is approximately normal with

mean of the sampling distribution of $\bar{x} = \mu = 13.6$

standard deviation of the sampling distribution of $\bar{x} = \frac{\sigma}{\sqrt{n}} = \frac{3.1}{\sqrt{22}} = 0.66$

Next, we note that from Table A the value of z, such that the area to the left of it under a standard normal curve is 0.05, is $z = -1.65$. Thus,

$-1.65 = (L - 13.6)/0.66.$

Solving for L gives

$L = (-1.65)(0.66) + 13.6 = 12.51$

SELECTED TEXT REVIEW EXERCISES

GUIDED SOLUTIONS

Exercise 9.42

KEY CONCEPTS: Sampling distributions for proportions

a) The flyer is sent to n = 25,000 Americans from 18 to 24 years old from a population in which p = 0.141 are high school dropouts. What is the mean number of high school dropouts who will receive the flyer? (You may want to think about this as a binomial distribution as described in Chapter 8.)

b) The normal approximation is given for the sampling distribution of a proportion with a particular characteristic, not the number with the characteristic. First, you should restate this event in terms of the proportion, $\hat{p}$, of dropouts in the sample who receive the flyer.

Event of interest in terms of $\hat{p}$ -

Next, you can approximate the probability using the normal approximation. What is the mean and standard deviation of the sampling distribution?

COMPLETE SOLUTIONS

Exercise 9.42

a) Those receiving the flyer are either dropouts (success) or not dropouts. The success probability is given as $p = 0.141$, and the number of observations is $n =$ 25,000. The mean for the number of successes (dropouts) in the sample is $np =$ (25000)(0.141) = 3525.

b) The event that at least 3500 dropouts receive the flyer is the same as the event that that at least $\hat{p} = 3500/25000 = 0.14$ of the sample receive the flyer. The probability for the event of interest is $P(\hat{p} \geq 0.14)$. For the sampling distribution of $\hat{p}$, the mean is $p = 0.141$, and the standard deviation is

$$\sqrt{\frac{p(1-p)}{n}} = \sqrt{\frac{0.141(1-0.141)}{25000}} = 0.0022$$

Using the normal approximation leads to

$$P(\hat{p} \geq 0.14) = P\left(\frac{\hat{p}-0.141}{0.0022} \geq \frac{0.14-0.141}{0.0022}\right) = P(Z \geq -.45) = 0.6736.$$

CHAPTER 10

INTRODUCTION TO INFERENCE

SECTION 10.1

OVERVIEW

A **confidence interval** provides an estimate of an unknown parameter of a population or process, along with an indication of how accurate this estimate is and how confident we are that the interval is correct. Confidence intervals have two parts. One is an interval computed from our data. This interval typically has the form

estimate ± margin of error

The other part is the **confidence level**, which states the probability that the *method* used to construct the interval will give a correct answer. For example, if you use a 95% confidence interval repeatedly, in the long run 95% of the intervals you construct will contain the correct parameter value. Of course, when you apply the method only once, you do not know whether or not your interval gives a correct value. Confidence refers to the probability that the method gives a correct answer in repeated use, not the correctness of any particular interval we compute from data.

Suppose we wish to estimate the unknown mean μ of a normal population with known standard deviation σ based on an SRS of size n. A level C confidence interval for μ is

$$\bar{x} \pm z^* \frac{\sigma}{\sqrt{n}}$$

in which z^* is such that the probability is C that a standard normal random variable lies between $-z^*$ and z^* and is obtained from the bottom row in Table C. These z-values are called critical values.

The margin of error $z^* \frac{\sigma}{\sqrt{n}}$ of a confidence interval decreases when any of the following occur:

- The confidence level C decreases.
- The sample size n increases.
- The population standard deviation σ decreases.

The sample size needed to obtain a confidence interval for a normal mean of the form

estimate ± margin of error

with a specified margin of error m is determined by solving the inequality

$$z^* \frac{\sigma}{\sqrt{n}} \leq m$$

where z^* is the critical value for the desired level of confidence. Many times, the n you find will not be an integer. If it is not, round up to the next larger integer.

The formula for any specific confidence interval is a recipe that is correct under specific conditions. The most important conditions concern the methods used to produce the data. Many methods (including those discussed in this section) assume that our data were collected by random sampling. Other conditions, such as the actual distribution of the population, also are important.

GUIDED SOLUTIONS

Exercise 10.1

KEY CONCEPTS: Confidence intervals, interpreting statistical confidence

a) The general form of a confidence interval is

estimate ± margin of error

Identify the estimate (percentage of the women who said they do not get enough time for themselves) and the margin of error. Combine these in the general form of a confidence interval as indicated.

b) In formulating your explanation, remember the notions of sampling distributions and sampling variability as discussed in Chapter 9. Write your explanation.

c) In formulating your explanation, consider the meaning of statistical confidence as described in Section 10.1 of the textbook or the overview for this section of the Study Guide. Write your explanation.

Exercise 10.12

KEY CONCEPTS: Confidence intervals for means, the effect of sample size on the margin of error

a) First, identify the following quantities. You will want to use the bottom row of Table C to find z^*.

$\bar{x} =$

$\sigma =$

$n =$

z^* (for a 95% confidence interval) =

Next, use the formula for a 95% confidence interval to compute the desired interval

$$\bar{x} \pm z^* \frac{\sigma}{\sqrt{n}} =$$

b) Use the same formula as in part a, but now with $n = 250$ rather than 1077.

$$\bar{x} \pm z^* \frac{\sigma}{\sqrt{n}} =$$

c) Use the same formula as in part a, but now with $n = 4000$.

$$\bar{x} \pm z^* \frac{\sigma}{\sqrt{n}} =$$

d) The margins of errors are the quantities after the $\pm$. You computed these in parts a, b, and c. List them here.

Margin of error for $n = 250$:

Margin of error for $n = 1077$:

Margin of error for $n = 4000$:

What pattern do you observe?

Exercise 10.13

KEY CONCEPTS: Confidence intervals, determining sample size

a) First, identify the following quantities. You will want to use the bottom row of Table C to find z^*.

$\bar{x} =$

$\sigma =$

$n =$

z^* (for a 98% confidence interval) =

Next, use the formula for a 98% confidence interval to compute the desired interval

$$\bar{x} \pm z^* \frac{\sigma}{\sqrt{n}} =$$

b) To find the sample size, use the expression

$$z^* \frac{\sigma}{\sqrt{n}} \leq m$$

and solve for n. Refer to part a for z^* and σ, and notice here $m = 0.0001$.

Exercise 10.17

KEY CONCEPTS: Interpreting confidence intervals

a) Review the meaning of statistical confidence and then restate this in plain language.

b) Does the 95% confidence interval contain 50%? What does this imply?

c) Recall the meaning of probability as discussed in Chapter 6. Does probability apply to the fixed unknown proportion of voters that prefer Carter? Explain.

Exercise 10.21

KEY CONCEPTS: Confidence intervals for means, interpreting statistical confidence

a) Read the statement of the problem carefully and complete the following.

The population about which the authors of the study want to draw conclusions is:

The population we can be certain about which the authors can draw conclusions is:

b) To construct the desired intervals, use the formula

$$\bar{x} \pm z^* \frac{\sigma}{\sqrt{n}}$$

but replace σ with s. What is s here? Write the confidence intervals in the spaces indicated below.

Food stores: $\bar{x} \pm z^* \frac{s}{\sqrt{n}} =$

Mass merchandisers: $\bar{x} \pm z^* \frac{s}{\sqrt{n}} =$

Pharmacies: $\bar{x} \pm z^* \frac{s}{\sqrt{n}} =$

c) In answering this question, consider whether the intervals you computed in part a overlap. Also, bear in mind your answer to part a (To what population do the results apply and to what population do you think the authors want the results to apply?).

COMPLETE SOLUTIONS

Exercise 10.1

a) The estimate of interest here (percentage of the women who said they do not get enough time for themselves) is 47%. Because the margin of error for a 95% confidence interval is ± 3%, the 95% confidence interval for the percent of all adult women who think they do not get enough time for themselves is 47% ± 3%, or from 44% to 50%.

b) The value 47% is based on a sample of 1025 women selected at random from all women in the United States (excluding Alaska and Hawaii). Another sample of 1025 women might yield a different percent. Repeated random samples of 1025 women will yield a variety of percents. These values will vary around the true percent of women in the United States (excluding Alaska and Hawaii) who feel they do not get enough time for themselves. No particular sample, however, will necessarily give the true value of this percent. Thus, we cannot be sure that 47% is the true percent of women in the United States (excluding Alaska and Hawaii) who feel they do not get enough time for themselves. All we can say is that 47% is likely to be "close" to the true percent.

c) Suppose we take all possible random samples of 1025 women. In each sample, suppose we determine the percent of women who think they do not get enough time for themselves. For each of these percents, suppose we add and subtract the margin of error for a 95% confidence interval. Of the resulting intervals, 95% will contain the actual percent of all adult women in the United States (excluding Alaska and Hawaii) who think they do not get enough time for themselves. This is what we mean by "95% confidence." Note that we do not know whether any particular interval (such as the 44% to 50% interval in part a) contains the true value of the percent. The confidence level of 95% refers only to the percent of the intervals produced by all samples that will contain the true percent.

Exercise 10.12

a) We are given that the sample mean $\bar{x}$ is 275, the standard deviation σ is 60, and the sample size n is 1077. Thus, a 95% confidence interval for the mean score μ in the population of all young women is

$$\bar{x} \pm z^* \frac{\sigma}{\sqrt{n}} = 275 \pm 1.96 \frac{60}{\sqrt{1077}} = 275 \pm 3.58$$

or 271.42 to 278.58.

b) We simply replace $n = 1077$ with $n = 250$ in our calculations and get

$$\bar{x} \pm z^* \frac{\sigma}{\sqrt{n}} = 275 \pm 1.96 \frac{60}{\sqrt{250}} = 275 \pm 7.44$$

or 267.56 to 282.44.

c) We use $n = 4000$ in our calculations and get

$$\bar{x} \pm z^* \frac{\sigma}{\sqrt{n}} = 275 \pm 1.96 \frac{60}{\sqrt{4000}} = 275 \pm 1.86$$

or 273.14 to 276.86.

d) Margin of error for $n = 250$: ± 7.44

Margin of error for $n = 1077$: ± 3.58

Margin of error for $n = 4000$: ± 1.86

We see that as the sample size increases, the margin of error decreases.

Exercise 10.13

a) We have

$\bar{x} = 10.0023$

$\sigma = 0.0002$

$n = 5$

z^* (for a 98% confidence interval) = 2.326

thus

$$\bar{x} \pm z^* \frac{\sigma}{\sqrt{n}} = 10.0023 \pm 2.326 \frac{0.0002}{\sqrt{5}} = 10.0023 \pm 0.000208$$

b) Using the values in part a and $m = 0.0001$, the expression

$$z^* \frac{\sigma}{\sqrt{n}} \leq m$$

becomes

$$2.326 \frac{0.0002}{\sqrt{n}} \leq 0.0001$$

or on rearrangement

$$2.326 \frac{0.0002}{0.0001} \leq \sqrt{n}$$

which becomes

$$4.652 \leq \sqrt{n}.$$

Squaring both sides gives

$$21.64 \leq n$$

so n should be at least 22.

Exercise 10.17

a) Suppose we listed all possible samples we could obtain by the sampling method used to get the sample actually taken. For each, suppose we calculated the percent of the sample that intended to vote for Carter and then attached the appropriate margin of error (the ± 2% for the sample in the article) to each of these percents. Ninety-five percent of the resulting intervals would contain the true percent of people who intended to vote for Carter. Notice that the 95% tells us the percent of samples collected identically to the one actually taken that would contain the true percent of voters favoring Carter. Any particular sample (such as the one actually taken of 51% ± 2%) either does or does not contain the true percent.

b) The 95% confidence interval for the true percent of voters favoring Carter was 51% ± 2% or (49%, 53%). This interval includes values below 50%, which does not rule out the possibility that less than 50% favor Carter. As a result, the polling organization had to say that the election was too close to call (using a 95% confidence level).

c) The true percent of the voters who favor Carter is not known, but it is a fixed value. This fixed value is either larger than 50% or it is not larger than 50%. There is no probability involved. Thus, the politician's question does not really make sense. Most likely, the politician is confusing a lack of knowledge, or uncertainty, about the true value with the notion of probability or chance.

Remember, probabilities in polls refer to the method of selecting a sample. In particular, they refer to whether the method used to select the sample will produce results within some given margin of error of the true value. Probabilities do not refer to the true value itself. Because the true value is unknown (that is why we are taking the sample in the first place), there is always a temptation to think of probabilities as referring to our lack of knowledge about this true value. In this textbook, avoid this temptation.

Note: Some statisticians define probability in terms of our lack of knowledge or uncertainty about unknown parameters of a population. This is not the view adopted by most statisticians or this textbook. If you take additional statistics courses, however, you may run across this alternate view. It is sometimes called the Bayesian or subjective view of probability.

Exercise 10.21

a) We would assume that the authors of the study wanted to draw conclusions about the population of all adult American consumers. Because the sample was drawn from the Indianapolis phone directory, we can be certain that they can draw conclusions only about the population of all people listed in this phone directory.

b) We use the formula $\bar{x} \pm z^* \frac{\sigma}{\sqrt{n}}$, but with the sample standard deviation, s, in place of σ. Here the sample size is $n = 201$, and because we want 95% confidence intervals, $z^* = 1.96$. The confidence intervals are

Food stores: $18.67 \pm 1.96\left(\frac{24.95}{\sqrt{201}}\right) = 18.67 \pm 3.45 = (15.22, 22.12)$

Mass merchandisers: $32.38 \pm 1.96\left(\frac{33.37}{\sqrt{201}}\right) = 32.38 \pm 4.61 = (27.77, 36.99)$

Pharmacies: $48.60 \pm 1.96\left(\frac{35.62}{\sqrt{201}}\right) = 48.60 \pm 4.92 = (43.68, 53.52)$

c) Neither of the other two 95% confidence intervals overlap with that for pharmacies. In fact, none is even close to overlapping with the interval for pharmacies. This seems reasonably strong evidence that consumers *from the population of all people listed in the Indianapolis phone directory* think pharmacies offer higher performance than other types of stores. Whether this is strong evidence that *all consumers* think pharmacies offer higher performance than other types of stores is less clear. We would need to decide to what extent the population of all people listed in the Indianapolis phone directory are representative of all consumers in the U.S. Without further information, we would be reluctant to extend our conclusions, or extrapolate, to the population of all U.S. consumers.

SECTION 10.2

OVERVIEW

Tests of significance and confidence intervals are the two most widely used types of formal statistical inference. A test of significance is done to assess the evidence against the **null hypothesis** H_0 in favor of an **alternative hypothesis** H_a. Typically, the alternative hypothesis is the effect that the researcher is trying to demonstrate, and the null hypothesis is a statement that the effect is not present. The alternative hypothesis can be either **one-** or **two-sided**.

Tests usually are carried out by first computing a **test statistic**. The test statistic is used to compute a ***P*-value**, which is the probability of getting a test statistic at least as extreme as the one observed, where the probability is computed when the null hypothesis is true. The P-value provides a measure of how incompatible our data are with the null hypothesis, or how unusual it would be to get data like ours if the null hypothesis were true. Because small P-values indicate data that are unusual or difficult to explain under the null hypothesis, we typically reject the null hypothesis in these cases. In this case, the alternative hypothesis provides a better explanation for our data.

Significance tests of the null hypothesis H_0: $\mu = \mu_0$ with either a one- or a two-sided alternative are based on the test statistic

$$z = \frac{\bar{x} - \mu_0}{\sigma / \sqrt{n}}$$

The use of this test statistic assumes that we have an SRS from a normal population with known standard deviation σ. When the sample size is large, the assumption of normality is less critical because the sampling distribution of $\bar{x}$ is approximately normal. P-values for the test based on z are computed using Table A.

When the P-value is below a specified value α, we say the results are **statistically significant at level** α, or we reject the null hypothesis at level α. Tests can be carried out at a fixed significance level by obtaining the appropriate critical value z^* from the bottom row in Table C.

GUIDED SOLUTIONS

Exercise 10.27

KEY CONCEPTS: Testing hypotheses about means, P-value

a) If the null hypothesis H_0: $\mu = 115$ is true, then scores in the population of older students are normally distributed, with mean $\mu = 115$ and standard deviation $\sigma = 30$. What then is the sampling distribution of $\bar{x}$, the mean of a sample of size $n = 25$? (We studied the sampling distribution of $\bar{x}$ in Chapter 9.)

Sketch the density curve of this distribution. Be sure to label the horizontal axis properly.

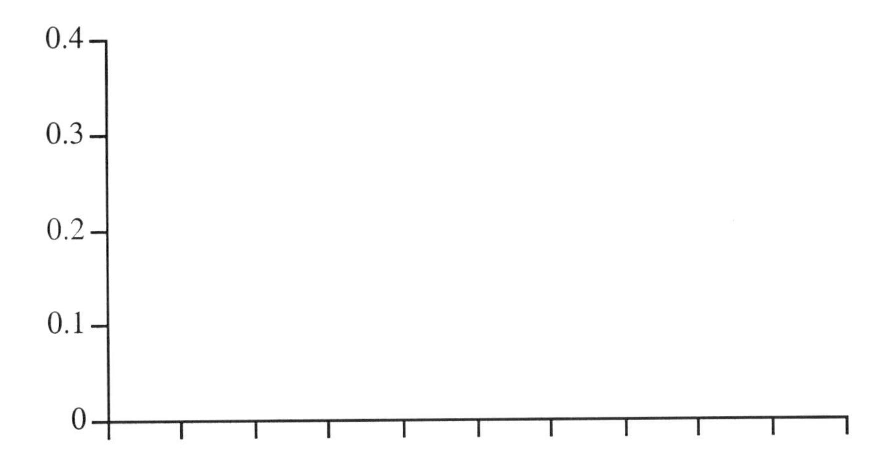

b) Mark the two points on your sketch in part a. Referring to this sketch, explain in simple language why one result is good evidence that the mean score of all older students is greater than 115 and why the other outcome is not. Think about how far out on the density curve the two points are.

c) Shade the appropriate area in your sketch in part a. Refer to Figure 10.9 in Chapter 10 of the textbook if you need a hint.

Exercise 10.31

KEY CONCEPTS: Null and alternative hypotheses

Remember that in many instances it is easier to begin with H_a, the effect that we hope to find evidence for, and then set up H_0 as the statement that the effect is not present. Think carefully about whether H_a should be one-sided or two-sided.

What does the teaching assistant hope to show? Use this to set up H_a and H_0.

Exercise 10.33

Key Concepts: P-values, statistical significance

Refer to the graph in your solution to exercise 10.27. The P-value for 118.6 is the shaded area, i.e., the area under the normal curve to the right of 118.6. We learned how to calculate such areas in Chapter 2. First you will need to find the z-score of 118.6, and then you will need to use Table A or your calculator to find the area to the right of this z-score under a standard normal curve. Use the space for your calculations.

Next, perform a similar calcualtion to find the P-value of 125.7.

Exercise 10.37

Key Concepts: Interpreting P-values

Write your explanation in the space. Refer to the section overview in this Study Guide or the first part of this section in the textbook if you need a hint.

Exercise 10.39

KEY CONCEPTS: Testing hypotheses, *P*-values.

a) Write the null and alternative hypotheses in the space provided. Remember, we are trying to decide whether the data give evidence that the process mean is not equal to the target value of 224 mm.

H_0: H_a:

b) Compute the test statistic. Recall that the formula is

$$z = \frac{\bar{x} - \mu_0}{\sigma / \sqrt{n}} =$$

c) Next, give the *P*-value. Use Table A or your calculator. Remember, the alternative is two-sided.

P-value =

Exercise 10.41

KEY CONCEPTS: Testing hypotheses at a fixed significance level

a) The z-test statistic for testing against a two-sided alternative, as in this problem, is $|z| = \left|\frac{\bar{x} - \mu_0}{\sigma / \sqrt{n}}\right|$. Identify μ_0, the standard deviation σ, the sample mean $\bar{x}$, and the sample size n. Then complete the calculation of the test statistic.

$$|z| = \left|\frac{\bar{x} - \mu_0}{\sigma / \sqrt{n}}\right| =$$

b) To answer, you will have to find the appropriate critical value from Table C. Note that we are testing against a two-sided alternative.

c) Follow the procedure in part b, but with significance level 1%.

Exercise 10.47

Key Concepts: Null and alternative hypotheses

Remember that in many instances it is easier to begin with H_a, the effect that we hope to find evidence for, and then set up H_0 as the statement that the effect is not present. Think carefully about whether H_a should be one-sided or two-sided.

The researcher hopes to show that on the average the mice complete the maze faster when there is a loud noise. What is H_a? What is H_0?

Exercise 10.52

Key Concepts: Calculating *P*-values

We need to compare the value $z = -1.37$ to the critical values in Table C. Because the test is two-sided, we use $|z| = 1.37$, and double the tail area. Its value is between the critical values $z^* = 1.282$ and $z^* = 1.645$. What two signficance levels do these critical values correspond to? Remember that the test is two-sided. What can we say about the *P*-value?

Next, use Table A to determine the *P*-value by computing the area to the left of -1.37. Is this area the *P*-value? Why or why not? If not, what do you need to do to find the *P*-value?

COMPLETE SOLUTIONS

Exercise 10.27

a) From Chapter 9, we know that the sampling distribution of $\bar{x}$ is normal with mean $\mu = 115$ and standard deviation $\sigma = 30/\sqrt{n} = 30/\sqrt{25} = 30/5 = 6$. A sketch of the density curve of this distribution follows.

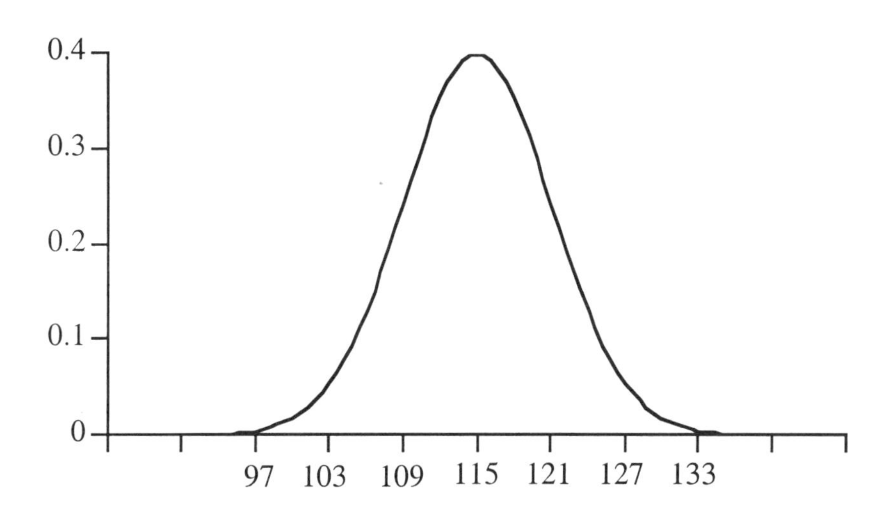

b) The two points are marked on the following curve.

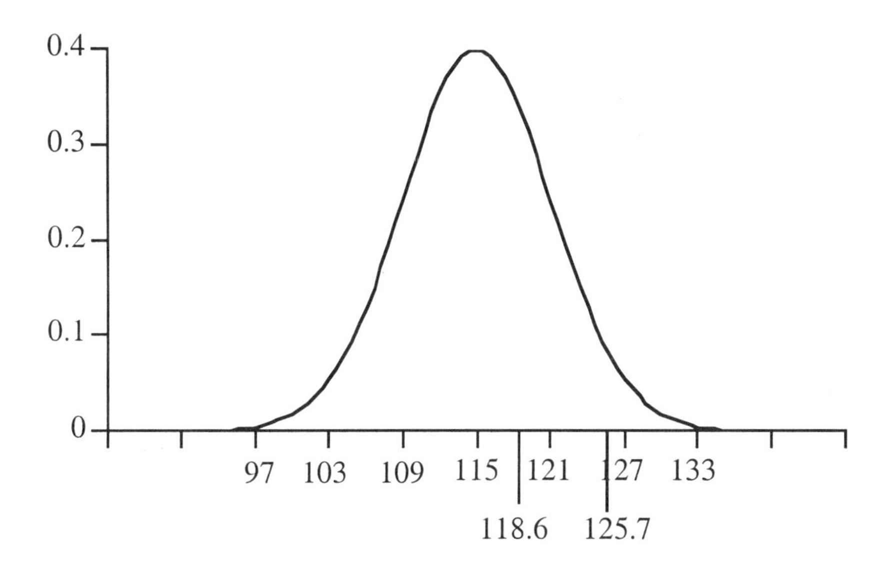

The 125.7 is much farther out on the normal curve than 118.6. In other words, it would be unlikely to observe a mean of 125.7 if the null hypothesis H_0: μ = 115 is true. However, a mean of 118.6 is fairly likely if the null hypothesis is true. A mean as large as 125.7 is more likely to occur if the true mean is larger than 115. Thus, 125.7 is good evidence that the mean score of all older students is greater than 115, while a mean score of 118.6 is not.

c) The P-value corresponds to the area to the right of 118.6, because the alternative hypothesis is H_a: $\mu > 115$. This area is displayed in the following figure.

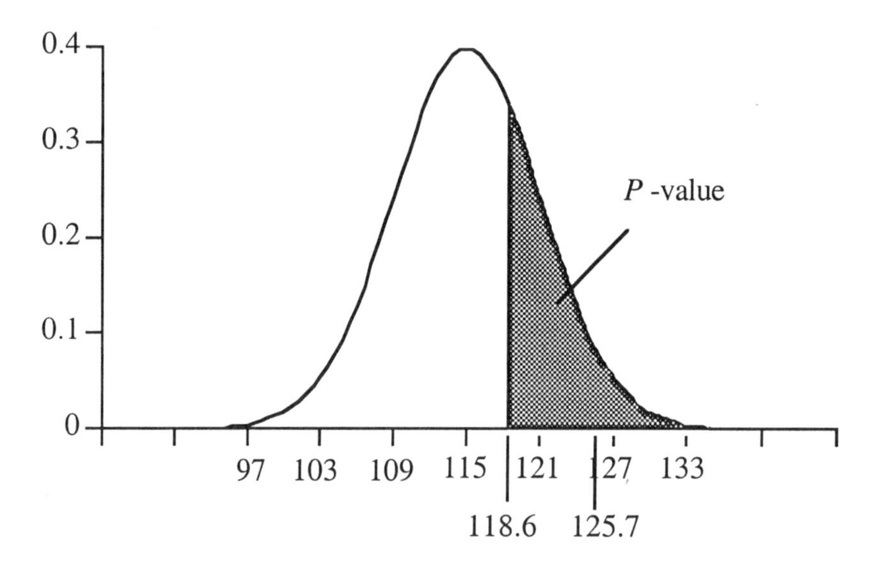

Exercise 10.31

The teaching assistant hopes to show that the students in classes he teaches will have a higher mean score than the class mean of 50. The hypotheses are H_0: μ = 50 and H_a: $\mu > 50$.

Exercise 10.33

Refer to the graph in the complete solution to Exercise 10.27. The P-value for 118.6 is the shaded area, i.e., the area under the normal curve to the right of 118.6. We learned how to calculate such areas in Chapter 2. First, we compute the z-score of 118.6. As we saw in exercise 10.27, the sampling distribution of $\bar{x}$ is normal with mean $\mu = 115$ and standard deviation $\sigma = 6$. We find that

$$z \text{ score} = \frac{118.6 - \mu}{\sigma} = \frac{118.6 - 115}{6} = \frac{3.6}{6} = 0.6$$

From Table A we find that the area under the standard normal curve to the right of 0.6 is 1 minus the area to the left of 0.6 = 1 – 0.7257 = 0.2743.

We make a similar calculation for the *P*-value of 125.7:

$$z \text{ score} = \frac{125.7-\mu}{\sigma} = \frac{125.7-115}{6} = \frac{10.7}{6} = 1.78$$

The area to the right of this z score under a standard normal curve = 1 minus the area to the left of 1.78 = 1 – 0.9625 = 0.0375.

In summary,

P-value of 118.6 = 0.2743

P-value of 125.7 = 0.0375

Exercise 10.37

In the sample selected by the psychologist, ethnocentrism among church attenders was higher than among nonattenders. Furthermore, the chance of obtaining a difference as large as that observed by the psychologist is less than 0.05 if, in fact, there is no real difference in the population from which the sample was selected. We would take this as strong evidence that ethnocentrism is higher among church attenders than among nonattenders in the population from which the sample was selected.

Exercise 10.39

a) The hypotheses are

$$H_0: \mu = 224 \qquad\qquad H_a: \mu \neq 224$$

The alternative is two-sided. We are trying to decide whether the data give evidence that the process mean is *not equal* to the target value of 224 mm.

b) Using a calculator, we find that the $n = 16$ measurements given have mean $\bar{x}$ = 224.001875. We are told that $\sigma = 0.060$ mm, so the test statistic is

$$z = \frac{\bar{x}-\mu_0}{\sigma/\sqrt{n}} = \frac{224.001875-224}{0.060/\sqrt{16}} = \frac{0.001875}{0.060/4} = \frac{0.001875}{0.015} = 0.125.$$

c) Because this is a two-sided test, the *P*-value is the sum of the areas above 0.125 and below –0.125 under a standard normal curve. We find (using either Table A or a calculator) that this area is (approximately)

P-value = 0.4 + 0.4 = 0.8.

The evidence that the process mean is not 224 is rather weak.

Exercise 10.41

a) Because the null hypothesis is H_0: $\mu = 0.5$, we have $\mu_0 = 0.5$; the standard deviation is $\sigma = 0.2887$, the sample mean is $\bar{x} = 0.4365$, and the sample size is $n = 100$; the value of the z-test statistic is

$$|z| = \left|\frac{\bar{x} - \mu_0}{\sigma / \sqrt{n}}\right| = \left|\frac{0.4365 - 0.5}{0.2887 / \sqrt{100}}\right| = \left|\frac{-0.0635}{0.02887}\right| = |-2.20| = 2.20$$

b) Because we are testing against a two-sided alternative, we need to find the upper $\alpha/2 = 0.05/2 = 0.025$ critical value in Table C. We see that this critical value is 1.96. Because our z-test statistic is greater than this critical value, the result is significant at the 0.05 level.

c) Next, we look for the upper $\alpha/2 = 0.01/2 = 0.005$ critical value in Table C. We see that this critical value is 2.576. Because our z-test statistic is smaller than this critical value, the result is not significant at the 0.01 level.

Exercise 10.47

The researcher hopes to show that on the average the mice complete the maze faster when there is a loud noise, namely that $\mu < 18$. Thus, the alternative hypothesis is H_a: $\mu < 18$ and the null hypothesis is H_0: $\mu = 18$, or it is equally correct to write the null hypothesis as H_0: $\mu \geq 18$.

Exercise 10.52

From Table C, $z^* = 1.282$ corresponds to a significance level of $2 \times 0.10 = 0.20$ because the test is two-sided. The tabled value $z^* = 1.645$ corresponds to a significance level of $2 \times 0.05 = 0.10$. We would reject at $\alpha = .20$ because $|z| = 1.37 > 1.282$, but not at $\alpha = 0.10$ because $|z| = 1.37 < 1.645$. So the P-value lies between 0.10 and 0.20.

Using Table A, the area to the left of -1.37 is 0.0853 and the P-value is $2 \times 0.0853 = 0.1706$ because the test is two-sided.

SECTION 10.3

OVERVIEW

When describing the outcome of a hypothesis test, it is more informative to give the *P*-value than to just report a decision to reject or not reject at a particular significance level α. The traditional levels of 0.01, 0.05, and 0.10 are arbitrary and serve as rough guidelines. Different people will insist on different levels of significance depending on the plausibility of the null hypothesis and the consequences of rejecting the null hypothesis. There is no sharp boundary between significant and insignificant, only increasingly strong evidence as the *P*-value decreases.

When testing hypotheses with a very large sample, the *P*-value can be rather small for effects that may not be of interest. Don't confuse small *P*-values with large or important effects. Statistical significance is not the same as practical significance. Plot the data to display the effect you are trying to show, and give a confidence interval that says something about the size of the effect.

Statistical inference from data based on a badly designed survey or experiment is often useless. Remember, a statistical test is valid only under certain conditions with data that have been properly produced.

Just because a test is not statistically significant doesn't imply that the null hypothesis is true. Statistical significance may occur when the test is based on a small sample size. Finally, if you run enough tests, invariably you will find statistical significance for one of them. Be careful at interpreting the results when testing many hypotheses on the same data.

GUIDED SOLUTIONS

Exercise 10.59

KEY CONCEPTS: Statistical significance versus practical importance

In this problem, we see that the *P*-value associated with the outcome $\bar{x}$ = 478 depends on the sample size. The probability of getting a value of $\bar{x}$ as large as 478 if the mean is 475 will become smaller as the sample size gets larger. (Do you remember why? Look at the formula for the *z*-score of $\bar{x}$.) Because this probability is the *P*-value, we see that a small effect is more likely to be detected for larger sample sizes than for smaller sample sizes. But this doesn't necessarily make the effect interesting or important. A confidence interval tells you something about the size of the effect, not the *P*-value.

a) Find the P-value by computing the z-test statistic and the probability of exceeding it.

b) This is the same as part a but with a larger sample size. The larger sample size makes the probability of getting a value of $\bar{x}$ as large as 478 smaller than it was in part a. Compute the P-value.

c) Find the P-value in this last case. It will be the smallest. Why?

Exercise 10.61

KEY CONCEPTS: Statistical inference is not valid for all sets of data

How was the survey conducted? Would statistical inference be valid for such a survey?

Exercise 10.62

KEY CONCEPTS: Multiple analyses

a) What does a P-value less than 0.01 mean? Out of 500 subjects, how many would you expect to achieve a score that has such a P-value if all 500 are guessing?

b) What would you suggest the researcher now do to test whether any of these four subjects have ESP?

COMPLETE SOLUTIONS

Exercise 10.59

See the guided solutions for a full explanation of the way sample size can change your *P*-values.

a) The z-test statistic is

$$z = \frac{\bar{x} - \mu_0}{\sigma / \sqrt{n}} = \frac{478 - 475}{100 / \sqrt{100}} = 0.3$$

and

$$P\text{-value} = P(Z > 0.3) = 1 - 0.6179 = 0.3821$$

because the alternative is one-sided.

b) The test statistic is

$$z = \frac{\bar{x} - \mu_0}{\sigma / \sqrt{n}} = \frac{478 - 475}{100 / \sqrt{1000}} = 0.95$$

and

$$P\text{-value} = P(Z > 0.95) = 1 - 0.8289 = 0.1711$$

c) The test statistic is

$$z = \frac{\bar{x} - \mu_0}{\sigma / \sqrt{n}} = \frac{478 - 475}{100 / \sqrt{10000}} = 3$$

and

$$P\text{-value} = P(Z > 3) = 1 - 0.9987 = 0.0013$$

Exercise 10.61

This was a call-in poll. The data were not collected by a well-designed survey using random sampling. As a result, the confidence interval is probably useless for making inferences about the population of all viewers.

Exercise 10.62

a) A P-value of 0.01 means that the probability a subject would do as well by merely guessing is only 0.01. Among 500 subjects, all of whom are merely guessing, we would therefore expect 1%, or 5, of them to do significantly better than random guessing ($P < 0.01$). Thus, in 500 tests it is not unusual to see four results with P-values on the order of 0.01, even if all are guessing and none have ESP.

b) Only these four subjects should be retested with a new, well-designed test. If all four again have low P-values (say, below 0.01 or 0.05), we have real evidence that they are not merely guessing. In fact, if any one of the subjects has a very low P-value (say, below 0.01), it would also be reasonably compelling evidence that the individual is not merely guessing. However, a single P-value on the order of 0.10 would not be particularly convincing.

SECTION 10.4

OVERVIEW

From the point of view of making decisions, H_0 and H_a are just two statements of equal status that we must decide between. One chooses a rule for deciding between H_0 and H_a on the basis of the probabilities of the two types of errors we can make. A **Type I error** occurs if H_0 is rejected when it is in fact true. A **Type II error** occurs if H_0 is accepted when in fact H_a is true. There is a clear relation between α-level significance tests and testing from the decision-making point of view. α is the probability of a Type I error.

To compute the Type II error probability of a significance test about a mean of a normal population:

- Write the rule for accepting the null hypothesis in terms of $\bar{x}$.

- Calculate the probability of accepting the null hypothesis when the alternative is true.

The **power** of a significance test is always calculated at a specific alternative hypothesis and is the probability that the test will reject H_0 when that alternative is true. The power of a test against any particular alternative is 1 minus the probability of a Type II error. Power is usually interpreted as the ability of a test to detect an alternative hypothesis or as the sensitivity of a test to an alternative hypothesis. The power of a test can be increased by increasing the sample size when the significance level remains fixed.

GUIDED SOLUTIONS

Exercise 10.66

KEY CONCEPTS: Type I and Type II error probabilities

a) Write the two hypotheses. Remember, we usually take the null hypothesis to be the statement of "no effect."

H_0:

H_a:

Describe the two types of errors as "false positive" and "false negative" test results.

b) Which error probability would you choose to make smaller (at the expense of making the other error probability larger) and why?

Exercise 10.68

KEY CONCEPTS: Type I and Type II error probabilities

a) If $\mu = 0$, what is the sampling distribution of $\bar{x}$? Use this to compute the probability the test rejects, i.e., the probability $\bar{x} > 0$.

b) If $\mu = 0.3$, what is the sampling distribution of $\bar{x}$? Use this to compute the probability the test accepts H_0, i.e., the probability $\bar{x} \leq 0$.

c) If $\mu = 1$, what is the sampling distribution of $\bar{x}$? Use this to compute the probability the test accepts H_0, i.e., the probability $\bar{x} \le 0$.

Exercise 10.70

KEY CONCEPTS: Power

Begin by rewriting the rule for rejecting H_0 in terms of $\bar{x}$. We help you by getting you started. The rule is to reject H_0 if $z \le -1.645$, or

$$z = \frac{\bar{x} - 300}{3/\sqrt{6}} = \frac{\bar{x} - 300}{1.22} \le -1.645.$$

What does this inequality imply about the values of $\bar{x}$?

a) If $\mu = 299$, what is the sampling distribution of $\bar{x}$? Use this to compute the probability that the test rejects, i.e., the probability that $\bar{x}$ takes on values (which you computed above) that lead to rejecting H_0 when the particular alternative $\mu = 299$ is true.

b) If $\mu = 295$, what is the sampling distribution of $\bar{x}$? Use this to compute the probability that the test rejects, i.e., the probability that $\bar{x}$ takes on values (which you computed above) that lead to rejecting H_0 when the particular alternative $\mu = 295$ is true.

c) What do you notice in parts a and b about the change in power as the true value of μ changes from 299 to 295?

Exercise 10.74

KEY CONCEPTS: Power and its relationship with the Type II error probability

What is the relationship between the probability of a Type I error and the level of significance?

What is the relationship between the power of a test at a particular alternative and the Type II error at this alternative? Use the value of the power that you calculated in Exercise 10.70 to compute the Type II error probability.

COMPLETE SOLUTIONS

Exercise 10.66

a) The two hypotheses are

H_0: the patient has no medical problem
H_a: the patient has a medical problem

One possible error is to decide

H_a: the patient has a medical problem

when, in fact, the patient does not really have a medical problem. This is a Type I error, which could be called a false positive in this setting. The other type of error is to decide

H_0: the patient has no medical problem

when, in fact, the patient does have a problem. This is a Type II error, which could be called a false negative in this setting.

b) Most likely we would choose to decrease the error probability for a Type II error, or the false negative probability. Failure to detect a problem (particularly a major problem) when one is present could result in serious consequences (such as death). While a false positive also can have serious consequences (painful or expensive treatment that is not necessary), it is not likely to lead to the kinds of consequences that a false negative could produce. For example, consider the consequences of failure to detect a heart attack, the presence of AIDS, or the presence of cancer. Note: There are cases in which some might argue that a false positive would be a more serious error than a false negative. For example, a false positive in a test for Down's Syndrome or a birth defect in an unborn baby might lead parents to consider an abortion. Some would consider this a much more serious error than to give birth to a child with a birth defect.

Exercise 10.68

a) If $\mu = 0$, the sampling distribution of $\bar{x}$ is normal with mean $\mu = 0$ and standard deviation $\frac{\sigma}{\sqrt{n}} = \frac{1}{\sqrt{9}} = 0.33$. Thus, the probability of a Type I error is the probability that $\bar{x} > 0$ when the null hypothesis is true. Computing the z score for $\bar{x}$, we get

$$P(\bar{x} > 0) = P\left(\frac{\bar{x} - 0}{0.33} > \frac{0 - 0}{0.33}\right) = P(Z > 0) = 0.5$$

b) If $\mu = 0.3$, the sampling distribution of $\bar{x}$ is normal with mean $\mu = 0.3$ and standard deviation $\frac{\sigma}{\sqrt{n}} = \frac{1}{\sqrt{9}} = 0.33$. We accept H_0 if $\bar{x} \leq 0$. Thus the probability of a Type II error when $\mu = 0.3$ is

$$P(\bar{x} \leq 0) = P\left(\frac{\bar{x} - 0.3}{0.33} \leq \frac{0 - 0.3}{0.33}\right) = P(Z \leq -0.91) = 0.1814$$

c) If $\mu = 1$, the sampling distribution of $\bar{x}$ is normal with mean $\mu = 1$ and standard deviation $\frac{\sigma}{\sqrt{n}} = \frac{1}{\sqrt{9}} = 0.33$. We accept H_0 if $\bar{x} \leq 0$. Thus, the probability of a Type II error when $\mu = 1$ is

$$P(\bar{x} \leq 0) = P\left(\frac{\bar{x} - 1}{0.33} \leq \frac{0 - 1}{0.33}\right) = P(Z \leq -3.0) = 0.0013$$

Exercise 10.70

We begin by rewriting the rule for rejecting H_0 in terms of $\bar{x}$. The rule is to reject H_0 if $z \le -1.645$, or

$$\frac{\bar{x} - 300}{3/\sqrt{6}} = \frac{\bar{x} - 300}{1.22} \le -1.645.$$

Rewriting this in terms of $\bar{x}$ gives the following rule for rejection. Reject H_0 if

$$\bar{x} \le (1.22)(-1.645) + 300 = 297.99$$

a) If $\mu = 299$, the sampling distribution of $\bar{x}$ is normal with mean 299 and standard deviation $\frac{3}{\sqrt{6}} = 1.22$. The power is the probability of rejecting H_0 when the particular alternative $\mu = 299$ is true. This probability is

$$\text{Power} = P(\bar{x} \le 297.99) = P\left(\frac{\bar{x} - 299}{1.22} \le \frac{297.99 - 299}{1.22}\right)$$

$$= P(Z \le -0.83) = 0.2033$$

b) If $\mu = 295$, the sampling distribution of $\bar{x}$ is normal with mean 295 and standard deviation $\frac{3}{\sqrt{6}} = 1.22$. The power is the probability of rejecting H_0 when the particular alternative $\mu = 295$ is true. This probability is

$$\text{Power} = P(\bar{x} \le 297.99) = P\left(\frac{\bar{x} - 295}{1.22} \le \frac{297.99 - 295}{1.22}\right)$$

$$= P(Z \le 2.45) = 0.9929$$

c) The power will be greater than in part b. We notice in parts a and b that as μ decreased, the power increased. Thus, we would suspect that the power will be greater when μ decreases further to 290. A more precise argument is the following. If $\mu = 290$, $\bar{x}$ is likely to be close to 290 (within a standard deviation or two of 290). 297.99 is several multiples of the standard deviation of 1.22 above 290. Thus, it is almost certain that $\bar{x}$ will be less than 297.99 and more likely to be less than 297.99 than if μ were 295 (which is closer to 297.99 than 290 is). This means the power will be very close to 1 and greater than in part b.

Exercise 10.74

Recall that the test in Exercise 10.70 used a significance level of 5% to test the hypotheses

$$H_0: \mu = 300$$
$$H_a: \mu < 300$$

The probability of a Type I error is the same as the significance level and hence is 0.05.

The probability of a Type II error at the alternative $\mu = 295$ is the probability of accepting the null hypothesis H_0: $\mu = 300$. One minus this probability is the probability of (correctly) rejecting the null hypothesis H_0: $\mu = 300$ when the alternative $\mu = 295$ is true. This last probability is the power at the alternative $\mu = 295$. We found this to be 0.9929 in part b of exercise 6.67. One minus this value is thus the Type II error. Hence the Type II error is $1 - 0.9929 = 0.0071$.

SELECTED CHAPTER REVIEW EXERCISES

GUIDED SOLUTIONS

Exercise 10.80

KEY CONCEPTS: Null and alternative hypotheses, carrying out a significance test about a mean

a) What do the researchers hope to show? This will be the alternative. Write down the null and alternative hypotheses.

b) The first step in carrying out the test is to compute the test statistic $z = \dfrac{\bar{x} - \mu_0}{\sigma / \sqrt{n}}$ which measures how far the sample mean is from the hypothesized value μ_0. To find the numerical value of z, you need to determine μ_0, σ, and n from the problem, and then compute $\bar{x}$, the mean DRP score from the data given.

$z =$

Once the value of the test statistic z has been determined, the P-value can be computed. The P-value is the probability that the test statistic takes a value at least as extreme as the one observed. In the space provided, write this down as a probability in terms of Z, the standard normal, and then use Table A to evaluate this probability.

P-value = $P(Z \qquad) =$

If you are having trouble doing this directly from the meaning of the P-value, refer to the rules for computing P-values given in the box in the textbook and try to understand the rationale behind them.

Next, try to interpret your result in plain language.

Exercise 10.84

KEY CONCEPTS: Interpreting P-values

Write your explanation in the space provided. Refer to the section overview in the Study Guide or the first part of Section 10.2 in the textbook if you need to refresh your memory as to the proper interpretation of a P-value.

COMPLETE SOLUTIONS

Exercise 10.80

a) The researchers hope to show that their district mean exceeds the national mean of 32. The alternative is $\mu > 32$, so the hypotheses of interest to the researchers are H_0: $\mu = 32$ and H_a: $\mu > 32$.

b) The value of the test statistic is $z = \dfrac{\bar{x} - \mu_0}{\sigma / \sqrt{n}} = \dfrac{35.091 - 32}{11 / \sqrt{44}} = 1.86$. Because the alternative is $\mu > 32$, the P-value is the chance of getting a value of $\bar{x}$ at least as large as 35.091 if the true mean were 32. In terms of the test statistic, this is equivalent to computing $P(Z \geq z) = P(Z \geq 1.86) = 1 - 0.9686 = 0.0314$.

There is evidence that the mean score of all third graders in this district exceeds the national mean of 32.

Exercise 10.84

This is not a correct explanation. Remember, the null hypothesis is either true or false. There is no probability involved. Probability enters when we talk about the possible values we will obtain from a sample that will be used to test the null hypothesis. Probability refers to the chance our sample will give us reliable or misleading information. A correct explanation of the P-value of .03 is that if the null hypothesis is true, there is a probability of only 3% that we would observe a sample result as inconsistent with the null hypothesis as actually observed by chance. In other words, P-values tell us the probability that our sample results are due to chance (assuming that the null hypothesis is true) as opposed to being the result of a real effect.

CHAPTER 11

INFERENCE FOR DISTRIBUTIONS

SECTION 11.1

OVERVIEW

Confidence intervals and significance tests for the mean μ of a normal population are based on the sample mean $\bar{x}$ of an SRS. When the sample size n is large, the central limit theorem suggests that these procedures are approximately correct for other population distributions. In Chapter 10, we considered the (unrealistic) situation in which we knew the population standard deviation, σ. In this section, we consider the more realistic case in which σ is not known and we must estimate σ from our SRS by the sample standard deviation s. In Chapter 10, we used the **one-sample z statistic**

$$z = \frac{\bar{x} - \mu}{\sigma/\sqrt{n}}$$

which has the $N(0,1)$ distribution. Replacing σ by s, we now use the **one-sample t statistic**

$$t = \frac{\bar{x} - \mu}{s/\sqrt{n}}$$

which has the ***t* distribution** with $n - 1$ **degrees of freedom**.

For every positive value of k there is a t distribution with k degrees of freedom, denoted $t(k)$. All are symmetric, bell-shaped distributions, similar in shape to normal distributions but with greater spread. As k increases, $t(k)$ approaches the $N(0,1)$ distribution.

A level C **confidence interval for the mean** μ of a normal population when σ is unknown is

$$\bar{x} \pm t^* \frac{s}{\sqrt{n}}$$

where t^* is the upper $(1 - C)/2$ critical value of the $t(n - 1)$ distribution whose value can be found in Table C of the text or from statistical software. The one-sample t confidence interval has the form

$$\text{estimate } \pm t^* \text{SE}_{\text{estimate}},$$

where "SE" stands for **standard error**.

Significance tests of H_0: $\mu = \mu_0$ are based on the one-sample t statistic. P-values or fixed significance levels are computed from the $t(n-1)$ distribution using Table C or, more commonly in practice, using statistical software.

One application of these one-sample t procedures is to the analysis of data from **matched pairs** studies. We compute the differences between the two values of a matched pair (often before and after measurements on the same unit) to produce a single sample value. The sample mean and standard deviation of these differences are computed. Depending on whether we are interested in a confidence interval or a test of significance concerning the difference in the population means of matched pairs, we use either the one-sample confidence interval or the one-sample significance test based on the t statistic.

For larger sample sizes, the t procedures are fairly **robust** against nonnormal populations. As a rule of thumb, t procedures are useful for nonnormal data when $n \geq 15$, unless the data show outliers or strong skewness. For samples of size $n \geq 40$, t procedures can be used for even clearly skewed distributions. For smaller samples, it is a good idea to examine stemplots or histograms before you use the t procedures to check for outliers or skewness.

GUIDED SOLUTIONS

Exercise 11.6

KEY CONCEPTS: Critical values of the $t(k)$ distribution

a) For confidence intervals, the critical value t^* is the upper $(1 - C)/2$ critical value of the $t(n - 1)$ distribution. We must identify the confidence level C and the sample size n. Because we are interested in a 95% confidence interval, C is 0.95. Because we take 10 observations, $n = 10$. Thus, t^* is the upper $(1 - 0.95)/2 = 0.025$ critical value of the $t(10 - 1) = t(9)$ distribution. Next, turn to Table C and look under the column labeled 0.025 (the desired tail probability) and in the row labeled 9 (the degrees of freedom). What value do you find in the table?

b) This is just like part a. Identify C and n here and proceed as in part a

c) Again, this is just like part a. See whether you can do this on your own.

Exercise 11.7

KEY CONCEPTS: Critical values of the $t(k)$ distribution, P-values for t procedures

a) For confidence intervals or significance tests, the critical values for the one-sample t procedures use the $t(n-1)$ distribution. What are the degrees of freedom for the t statistic given in this problem?

b) Between what two critical values t^* does $t = 1.82$ fall? What are the right-tail probabilities corresponding to these entries? In which row of Table C do you need to look to answer these questions?

c) The P-value is $P(T \geq 1.82)$ because the alternative is H_a: $\mu > 0$. Between what two values does the P-value lie?

d) Is the value $t = 1.82$ significant at the 5% level? Why or why not?

What about at the 1% level?

Exercise 11.13

KEY CONCEPTS: Matched pairs experiments, one-sample t tests

a) This is a matched-pairs experiment. The matched pair of observations are the right- and left-hand times on each subject. To avoid confounding with time of day, we would probably want subjects to use both knobs in the same session. We also would want to randomize which knob the subject uses first. How might you do this randomization? What about the order in which the subjects are tested?

b) The project hopes to show that right-handed people find right-hand threads easier to use than left-hand threads. In terms of the mean μ for the population of differences

(left thread time) – (right thread time)

what do we wish to show? This would be the alternative. What are H_0 and H_a (in terms of μ)?

H_0:
H_a:

c) For data from a matched pairs study, we compute the differences between the two values of a matched pair to produce a single sample value. These differences are given below for our data.

Right thread	Left thread	Difference = Left – Right
113	137	24
105	105	0
130	133	3
101	108	7
138	115	–23
118	170	52
87	103	16
116	145	29
75	78	3

Right thread	Left thread	Difference = Left – Right (continued)
96	107	11
122	84	–38
103	148	45
116	147	31
107	87	–20
118	166	48
103	146	43
111	123	12
104	135	31
111	112	1
89	93	4
78	76	–2
100	116	16
89	78	–11
85	101	16
88	123	35

The sample mean and standard deviation of these differences need to be computed. Fill in their values in the space provided.

$\bar{x} =$ $s =$

Next we use the one sample significance test based on the t statistic. What value of μ_0 should be used?

$$t = \frac{\bar{x} - \mu_0}{s/\sqrt{n}} =$$

From the value of the t statistic and Table C (or using a calculator or statistical software), the P-value can be computed. If using Table C, between what two values does the P-value lie?

$\leq P\text{-value} \leq$

What conclusion do you draw?

Note: This problem is most easily done directly using statistical software or your graphing calculator. The software will compute the differences, the t statistic, and the P-value for you. Consult your users manual to see how to do one-sample t tests.

Exercise 11.17

a) Recall some of the graphical methods of Chapter 1 for describing data. We have drawn a boxplot below. Construct a histogram, either by hand using the class intervals provided, by calculator, or statistical software.

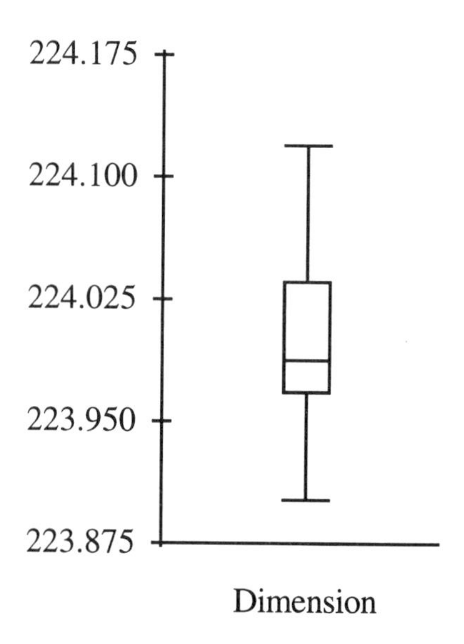

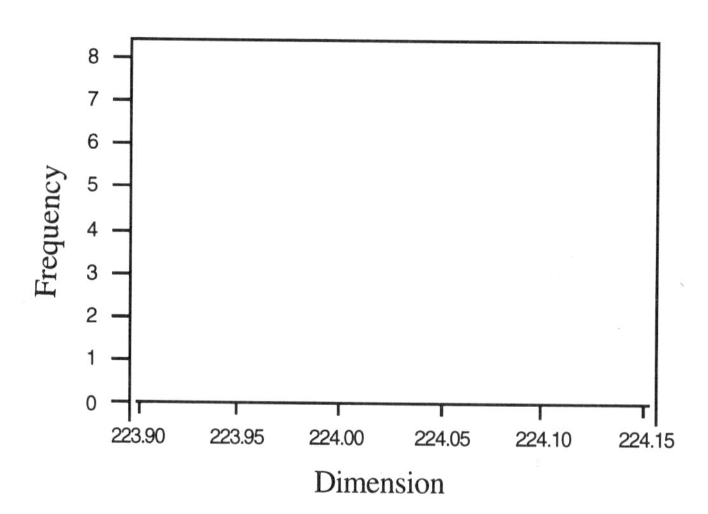

Using the histogram and/or the boxplot, is there evidence for outliers or strong skewness? Given the general guidelines on the robustness of the t procedures, do you think it is valid to use these procedures for these data?

b) We are interested in whether the data provide evidence that the mean dimension is not 224 mm. State the appropriate null and alternative hypotheses below.

H_0:
H_a:

From the data, first calculate the sample mean and the sample standard deviation, preferably using statistical software or a calculator. Fill in the values below.

$\bar{x} =$ $s =$

Next use these to compute the value of the one-sample t statistic.

$$t = \frac{\bar{x} - \mu_0}{s/\sqrt{n}} =$$

Compute the P-value using Table C or software. How many degrees of freedom are there? The degrees of freedom tell you which row of Table C you need to refer to for critical values. Remember, if the alternative is two-sided, then the probability found in the table needs to be doubled.

Degrees of freedom:

P-value:

Conclusion:

Exercise 11.20

KEY CONCEPTS: Power of the one-sample t test

a) You may wish to refer to Section 10.4 to review power calculations for the z test. To assist you, we give a complete solution. We wish to determine the power of the test against the alternative $\mu = 0.5$ when the significance level is 0.05 and $n = 10$. We use 0.83 as an estimate of both the population standard deviation σ and s in future samples. The t test with 10 observations rejects H_0: $\mu = 0$ (this is the null hypothesis that the mean difference in yields is 0) if the t statistic

$$t = \frac{\bar{x} - 0}{s/\sqrt{10}}$$

exceeds the upper 5% point of $t(9)$, which is 1.833 (obtained using Table C). Using $s = 0.83$, the event that the test rejects H_0 is

$$t = \frac{\bar{x}}{0.83/\sqrt{10}} \geq 1.833$$

or

$$\bar{x} \geq 1.833 \frac{0.83}{\sqrt{10}} = 0.481.$$

The power is the probability that $\bar{x} \geq 0.481$ when $\mu = 0.5$. Taking $\sigma = 0.83$, this probability is found by standardizing $\bar{x}$,

$$P(\bar{x} \geq 0.481 \text{ when } \mu = 0.5) = P\left(\frac{\bar{x} - 0.5}{0.83/\sqrt{10}} \geq \frac{0.481 - 0.5}{0.83/\sqrt{10}}\right)$$

$$= P(Z \geq -0.072) = 1 - 0.4713 = 0.5287.$$

Thus the power is 0.5287.

b) Try this one on your own. Use the same argument as in part a, but now with $n = 25$ rather than $n = 10$ wherever the value of n appears. All other features of the problem are the same.

Exercise 11.23

KEY CONCEPTS: Confidence intervals based on the one-sample t statistic, assumptions underlying t procedures

a) To compute a level C confidence interval, we use the formula $\bar{x} \pm t^* \frac{s}{\sqrt{n}}$ in which t^* is the upper $(1 - C)/2$ critical value of the $t(n - 1)$ distribution, which can be found in Table C. Fill in the missing values below. Don't forget to subtract one from the sample size when finding the appropriate degrees of freedom for the t confidence interval.

$C =$
$n =$
$t^* =$

The values of $\bar{x}$ and s are given in the problem.

$\bar{x} =$ $\qquad$ $s =$

Substitute all these values into the formula to complete the computation of the 95% confidence interval.

$$\bar{x} \pm t^* \frac{s}{\sqrt{n}} =$$

b) What are the assumptions required for the t confidence interval? Which assumptions are satisfied and which might not be? How were the subjects in the study obtained? How were the subjects in the placebo group obtained?

Exercise 11.29

KEY CONCEPTS: Appropriateness of statistical procedures

Statistical procedures use information in a sample to make inferences about parameters of a population. Ask yourself, what is the population and the parameter of interest in this exercise? What is the sample?

COMPLETE SOLUTIONS

Exercise 11.6

a) The value of t^* that you should find in Table C is 2.262.

b) Because we are interested in a 99% confidence interval, C is 0.99. Because we take an SRS of 20 observations, $n = 20$. The desired critical value t^* is the upper $(1 - 0.99)/2 = 0.005$ critical value of the $t(20 - 1) = t(19)$ distribution. Next, turn to Table C and look in the column labeled 0.005 and the row labeled 19. You should find that

$$t^* = 2.861.$$

c) Because we are interested in an 80% confidence interval, C is 0.80. Because we take a sample of size 7, $n = 7$. Thus t^* is the upper $(1 - 0.80)/2 = 0.10$ critical value of the $t(7 - 1) = t(6)$ distribution. Next, turn to Table C and look in the column labeled 0.10 and the row labeled 6. You should find that

$$t^* = 1.440.$$

Exercise 11.7

a) The degrees of freedom for the one-sample t procedures are $n - 1 = 15 - 1 = 14$.

b) You need to refer to the row corresponding to 14 degrees of freedom. The value of $t = 1.82$ falls between the two t^* critical values 1.761 and 2.145, corresponding to upper tail probabilities of 0.05 and 0.025, respectively.

df = 14

p	.05	.025
t^*	1.761	2.145

c) The $P(T \geq 1.82)$ must lie between 0.025 and 0.05. (If the alternative were two-sided, these values would need to be doubled.)

d) The value $t = 1.82$ is signficant at the 5% level because the P-value is below 0.05, but not at the 1% level because the P-value exceeds 0.01.

Exercise 11.13

a) The randomization might be carried out by simply flipping a fair coin. If the coin comes up heads, use the right-hand threaded knob first. If the coin comes up tails, use the left-hand threaded knob first. Alternatively, in order to balance out the number of times each type is used first, one might choose an SRS of 12 of the 25 subjects. These 12 use the right-hand thread knob first. Everyone else uses the left-hand thread knob first.

A second place one might use randomization is in the order in which subjects are tested. Use a table of random digits to determine this order. Label subjects 01 to 25. The first label that appears in the list of random digits (read in groups of two digits) is the first subject measured. The second label that appears, the next subject measured, etc. This randomization is probably less important than the one described in the previous paragraph. It would be important if the order or time at which a subject was tested might have an effect on the measured response. For example, if the study began early in the morning, the first subject might be sluggish if still sleepy. Sluggishness might lead to longer times and perhaps a larger difference in times. Subjects tested later in the day might be more alert.

b) In terms of μ, the mean of the population of differences, (left thread time) – (right thread time), we wish to test whether the times for the left-threaded knobs are longer than for the right-threaded knobs, i.e.,

$$H_0\text{: } \mu = 0 \text{ vs. } H_a\text{: } \mu > 0$$

c) For the 25 differences, we compute

$$\bar{x} = 13.32 \qquad s = 22.94.$$

We then use the one-sample significance test based on the t statistic.

$$t = \frac{\bar{x} - \mu_0}{s/\sqrt{n}} = \frac{13.32 - 0}{22.94/\sqrt{25}} = 2.903.$$

From the value of the t statistic and Table C, the P-value is between 0.0025 and 0.005.

df = 24

p	.005	.0025
t*	2.797	3.091

Using statistical software, the P-value is computed as P-value = 0.0039.

We conclude that there is strong evidence that the time for left-hand threads is greater than for right-hand threads on average.

Exercise 11.17

a) The boxplot was provided in the guided solutions and the completed histogram is given below (one could also make a stemplot of the data).

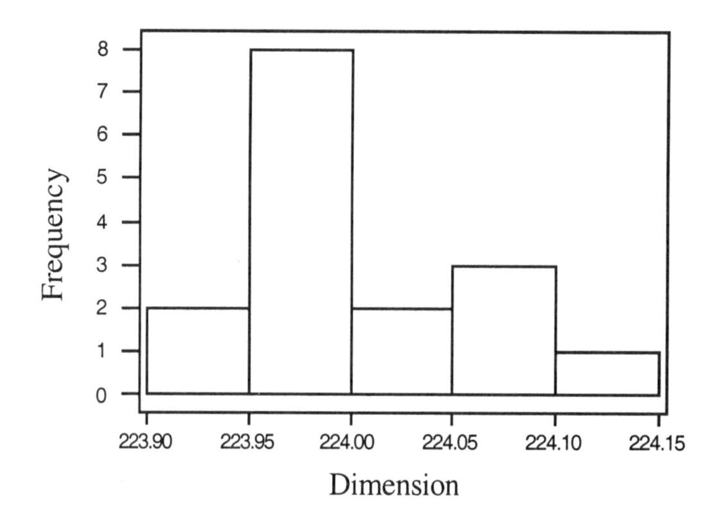

Using either the histogram or the boxplot, we can see that there are no outliers in the data. The data appear a bit skewed to the right, but not so strongly as to threaten the validity of the t procedure given that the sample size is 16. (In the section on the robustness of t procedures, t procedures are safe for samples of size $n \geq 15$ unless there are outliers and/or strong skewness.)

b) Because we are interested in whether the data provide evidence that the mean dimension is not 224 mm (no direction of the difference is specified), we wish to test the hypotheses

H_0: $\mu = 224$ mm

H_a: $\mu \neq 224$ mm

From the data, we calculate the basic statistics to be

$\bar{x} = 224.0019$, $s = 0.618$, and the standard error as $\dfrac{s}{\sqrt{n}} = \dfrac{0.0618}{\sqrt{16}} = 0.01545.$

Substituting these in the formula for t yields

$$t = \frac{\bar{x} - \mu_0}{s/\sqrt{n}} = \frac{224.0019 - 224}{0.01545} = 0.123.$$

The P-value for $t = 0.123$ is twice the area to the right of 0.123 under the t distribution curve with $n - 1 = 15$ degrees of freedom. Using Table C, we search the df = 15 row for entries that bracket 0.123. Because 0.123 lies to the left of the smallest entry in the Table corresponding to a probability of 0.25, the

df = 15

p	.25
t^*	0.691

P-value is therefore greater than .25 × 2 = .50 for this two-sided test. The data do not provide strong evidence that the mean differs from 224 mm. (Computer software gives a P-value of 0.9038.)

Exercise 11.20

a) A complete solution is given in the guided solution.

b) The t test with 25 observations rejects H_0: $\mu = 0$ (this is the null hypothesis that the mean difference in yields is 0) if the t statistic

$$t = \frac{\bar{x} - 0}{s/\sqrt{25}}$$

exceeds the upper 5% point of $t(24)$, which is 1.711. Using $s = 0.83$, the event that the test rejects H_0 is

$$t = \frac{\bar{x}}{0.83/\sqrt{25}} \geq 1.711$$

or

$$\bar{x} \geq 1.711\frac{0.83}{\sqrt{25}} = 0.284.$$

The power is the probability that $\bar{x} \geq 0.284$ when $\mu = 0.5$. Taking $\sigma = 0.83$, this probability is found by standardizing $\bar{x}$,

$$P(\bar{x} \geq 0.284 \text{ when } \mu = 0.5) \quad = P\left(\frac{\bar{x} - 0.5}{0.83/\sqrt{25}} \geq \frac{0.284 - 0.5}{0.83/\sqrt{25}}\right)$$

$$= P(Z \geq -1.301) = 1 - 0.0966 = 0.9034.$$

Thus the power is 0.9034.

Exercise 11.23

a) A 95% confidence interval for the mean systolic blood pressure in the population from which the subjects were recruited can be calculated from the data on the 27 members of the placebo group because these are randomly selected from the 54 subjects. We use the formula for a t interval, namely $\bar{x} \pm t^* \frac{s}{\sqrt{n}}$. In this problem, $\bar{x} = 114.9$, $s = 9.3$, $n = 27$, hence t^* is the upper $(1 - 0.95)/2 = 0.025$ critical value for the $t(26)$ distribution. From Table C, we see that $t^* = 2.056$. Thus, the 95% confidence interval is

$$114.9 \pm 2.056 \frac{9.3}{\sqrt{27}} = 114.9 \pm 3.68 = (111.22, 118.58)$$

b) For the procedure used in part a, the population from which the subjects were drawn should be such that the distribution of the seated systolic blood pressure in the population is normal. The 27 subjects used for the confidence interval in part a should be a random sample from this population. Unfortunately, we do not know if the latter is the case. While 27 subjects were selected at random from the total of 54 subjects in the study, we do not know whether the 54 subjects were a random sample from this population.

With a sample of 27 subjects, it is not crucial that the population be normal, as long as the distribution is not strongly skewed and the data contain no outliers. It is important that the 27 subjects can be considered a random sample from the population. If not, we cannot appeal to the central limit theorem to ensure that the t procedure is at least approximately correct even if the data are not normal.

(Note: It turns out that because the subjects were divided at random into treatment and control groups, there do exist procedures for comparing the treatment and placebo groups. These are not based on the t distribution, but are valid as long as treatment groups are determined by randomization. However, the conclusions drawn from these procedures apply only to the subjects in the study. To generalize the conclusions to a larger population, we must know that the subjects are a random sample from this larger population.)

Exercise 11.29

We do not have a sample of U.S. presidents. We have the ages of the entire *population* of U.S. presidents. The population average age is known exactly by simply averaging the numbers in Table 1.3. A confidence interval is not necessary because there is no uncertainty concerning the value of the population mean.

SECTION 11.2

OVERVIEW

One of the the most commonly used significance tests is the **comparison of two population means,** μ_1 and μ_2. In this setting, we have two distinct, independent SRSs from two populations or two treatments in a randomized comparative experiment. The procedures are based on the difference $\bar{x}_1 - \bar{x}_2$. When the populations are not normal, the results obtained using the methods of this section are approximately correct due to the central limit theorem.

Tests and confidence intervals for the difference in the population means, $\mu_1 - \mu_2$, are based on the **two-sample *t* statistic**. Despite the name, this test statistic does *not* have an exact t distribution. However, there are good approximations to its distribution, which allow us to carry out valid significance tests. Conservative procedures use the $t(k)$ distribution as an approximation in which the degrees of freedom k is taken to be the smaller of $n_1 - 1$ and $n_2 - 1$. More accurate procedures use the data to estimate the degrees of freedom k. This is the procedure that is followed by most statistical software and the TI-83 calculator.

To carry out a significance test for $H_0: \mu_1 = \mu_2$, use the two-sample t statistic

$$t = \frac{(\bar{x}_1 - \bar{x}_2)}{\sqrt{\frac{s_1^2}{n_1} + \frac{s_2^2}{n_2}}}$$

The P-value is found by using the approximate distribution $t(k)$, where k is estimated from the data when using statistical software or a calculator, or can be taken to be the smaller of $n_1 - 1$ and $n_2 - 1$ for a conservative procedure.

An approximate confidence C level **confidence interval** for $\mu_1 - \mu_2$ is given by

$$(\bar{x}_1 - \bar{x}_2) \pm t^* \sqrt{\frac{s_1^2}{n_1} + \frac{s_2^2}{n_2}}$$

in which t^* is the upper $(1 - C)/2$ critical value for $t(k)$, where k is estimated from the data when using statistical software, or can be taken to be the smaller of $n_1 - 1$ and $n_2 - 1$ for a conservative procedure. The procedures are most robust to failures in the assumptions when the sample sizes are equal.

The **pooled two-sample *t* procedures** are used when we can safely assume that the two populations have equal variances. The modifications in the procedure are the use of the pooled estimator of the common unknown variance and critical values obtained from the $t\,(n_1 + n_2 - 2)$ distribution.

GUIDED SOLUTIONS

Exercise 11.32

KEY CONCEPTS: Single sample, matched pairs or two-samples

a) Are there one or two samples involved? Was matching done?

b) Are there one or two samples involved? Was matching done?

Exercise 11.47

KEY CONCEPTS: Tests using the two-sample t, generalizations beyond the data

a) When means and standard deviations for the two samples are given, it is relatively easy to compute the value of the two-sample t statistic. When the original observations are given, the details of the computations are best left to statistical software. However, even when you let the calculator or computer do the tedious computations, it is still important to know what all the numbers in the output mean and how to interpret the results.

The solution to this problem is given in three parts. The first part leads you through the required computations and then uses the conservative degrees of freedom. The second part illustrates the use the more accurate approximation to the degrees of freedom. The third part shows you how to interpret the output that would be produced for this problem using the statistical software MINITAB.

The problem provides the summary statistics for the two samples that are required for computing the two-sample t statistic. For the Low group, $n_1 = 14$, $\bar{x}_1 = 4.640$, $s_1 = 0.690$. For the High group, $n_2 = 14$, $\bar{x}_2 = 6.429$, $s_2 = 0.430$.

The first step is to determine the null and alternative hypothesis. You need to decide whether the alternative is one-sided or two-sided from the question of interest described in the wording of the problem. Write down the appropriate null and alternative below.

H_0:
H_a:

The next step is to compute the value of the t statistic. Use the summary statistics provided to complete the calculations in the space provided.

$$t = \frac{\bar{x}_1 - \bar{x}_2}{\sqrt{\frac{s_1^2}{n_1} + \frac{s_2^2}{n_2}}} =$$

The conservative degrees of freedom uses the smaller of $n_1 - 1$ and $n_2 - 1$. After you have determined the degrees of freedom, compare the computed value of t to the values in Table C using the appropriate degrees of freedom. Is the difference in mean ego strength significant at the 5% level? At the 1% level? What are your conclusions?

Evaluate the approximate degrees of freedom using the formula

$$\frac{\left(\frac{s_1^2}{n_1} + \frac{s_2^2}{n_2}\right)^2}{\frac{1}{n_1 - 1}\left(\frac{s_1^2}{n_1}\right)^2 + \frac{1}{n_2 - 1}\left(\frac{s_2^2}{n_2}\right)^2} =$$

Is the difference in mean ego strength significant at the 5% level? At the 1% level? What are your conclusions?

The original data for this problem were entered into MINITAB. The software produced the output provided below.

Two Sample T-Test and Confidence Interval

```
Two-sample T for Low vs. High
        N      Mean      StDev    SE Mean
Low    14     4.640      0.690       0.18
High   14     6.429      0.430       0.12

T-Test mu Low = mu High (vs not =): T= -8.23 P=0.0000 DF = 21.77
```

The output begins with summary information for the two samples. For the Low group, $n_1 = 14$, $\bar{x}_1 = 4.640$, $s_1 = 0.690$ and the SE Mean is $s_1 / \sqrt{n_1} = 0.18$. For the High group, $n_2 = 14$, $\bar{x}_2 = 6.429$ $s_2 = 0.430$ and the SE Mean is $s_2 / \sqrt{n_2} =$ 0.12. The next line gives the hypotheses, test statistic and P-value:

`mu Low = mu High` corresponds to H_0: $\mu_{\text{Low}} = \mu_{\text{High}}$, and
`(vs not =)` corresponds to H_a: $\mu_{\text{Low}} \neq \mu_{\text{High}}$.

You should have verified the value of the $t = -8.23$ statistic previously, as well as the approximate df = 21.77. The P-value is also given. Use the information in the P-value to answer the questions: Is the difference in mean ego strength significant at the 5% level? At the 1% level? Why?

b) Are these groups random samples from low-fitness and high-fitness groups of middle-aged men? Where did the individuals in the samples come from? How might this affect generalizing to the population of all middle-aged men?

Exercise 11.51

KEY CONCEPTS: Tests and confidence intervals using the two-sample t, relationship between two-sided tests and confidence intervals

The summary statistics for the comparison of children and adult VOT scores are reproduced below.

Group	n	$\bar{x}$	s
Children	10	–3.67	33.89
Adults	20	–23.17	50.74

a) Were the researchers interested in a difference in either direction? State the appropriate null and alternative hypotheses in the space provided.

H_0:
H_a:

Next, find the numerical value of the two-sample t statistic using the formula below. Don't forget to square both standard deviations when evaluating the formula for the standard error in the denominator of the t.

$$t = \frac{\bar{x}_1 - \bar{x}_2}{\sqrt{\frac{s_1^2}{n_1} + \frac{s_2^2}{n_2}}} =$$

Compute the P-value using a t distribution with the smaller of $n_1 - 1 = 9$ and $n_2 - 1 = 19$ degrees of freedom, and then compare the computed value of t to the critical values given in Table C. Remember, if the test is two-sided you will need to double the tail probabilities. State your conclusions.

b) The formula for the 95% confidence interval is $\bar{x}_1 - \bar{x}_2 \pm t^* \sqrt{\frac{s_1^2}{n_1} + \frac{s_2^2}{n_2}}$, in which t^* is the upper $(1 - C)/2 = 0.025$ critical value for the t distribution with degrees of freedom equal to the smaller of $n_1 - 1 = 9$ and $n_2 - 1 = 19$. Find t^* from Table C and complete the computations for the confidence interval.

Remember the relationship between two-sided tests and confidence intervals. We reject the null hypothesis $H_0: \mu_1 = \mu_2$ at significance level α when the $1 - \alpha$ confidence interval for $\mu_1 - \mu_2$ doesn't contain the value 0. How did you know from your result in part a that the interval would contain 0 (no difference)?

COMPLETE SOLUTIONS

Exercise 11.32

a) This example involves a single sample. We have a sample of 20 measurements and we want to see whether the mean for this method agrees with the known concentration.

b) This example involves two samples, the set of measurements on each method. Note that we are not told of any matching.

Exercise 11.47

a) The hypotheses are

H_0: $\mu_{\text{Low}} = \mu_{\text{High}}$, H_a: $\mu_{\text{Low}} \neq \mu_{\text{High}}$

because the researchers were interested in a difference in either direction. The computations for the t statistic give

$$t = \frac{\bar{x}_1 - \bar{x}_2}{\sqrt{\frac{s_1^2}{n_1} + \frac{s_2^2}{n_2}}} = \frac{4.640 - 6.429}{\sqrt{\frac{(0.690)^2}{14} + \frac{(0.430)^2}{14}}} = -8.23.$$

Using the smaller of $n_1 - 1 = 13$ and $n_2 - 1 = 13$, the conservative degrees of freedom is 13, so the value of $t = -8.23$ must be compared with critical values from the line in Table C corresponding to 13 degrees of freedom. For testing at the 5% level, the critical value t^* for the two-sided alternative is 2.160. So, reject H_0 whenever $t > 2.160$ or $t < -2.160$. Because $t = -8.23$, we reject at the 5% level. Similiarly, at the 1% level $t^* = 3.012$ and we reject at the 1% level. There is strong evidence that the mean ego strength is higher for the high fitness group.

The formula for the approximate degrees of freedom gives

$$\frac{\left(\frac{s_1^2}{n_1} + \frac{s_2^2}{n_2}\right)^2}{\frac{1}{n_1 - 1}\left(\frac{s_1^2}{n_1}\right)^2 + \frac{1}{n_2 - 1}\left(\frac{s_2^2}{n_2}\right)^2} = \frac{\left(\frac{0.690^2}{14} + \frac{0.430^2}{14}\right)^2}{\frac{1}{14 - 1}\left(\frac{0.690^2}{14}\right)^2 + \frac{1}{14 - 1}\left(\frac{0.430^2}{14}\right)^2} = 21.77.$$

Going to the Table C under df = 21, we again reject at the 5% level of significance and the 1% level of significance. Of course, if we reject with the "conservative" degrees of freedom at a given significance level, we would also reject at this level using the approximate degrees of freedom computed from the data. That is the meaning of the word conservative in this context.

From the computer output, we are given the P-value as $P = 0.0000$. If we know the P-value, we can assess the significance at any level by comparing the P-value to the significance level. Because the P-value is the smallest α level at which we can reject H_0, if the P-value is smaller than the significance level given, we reject H_0; otherwise we do not reject. Because the P-value in the output is given as 0.0000, it is smaller than both 5% and 1%. So we reject H_0 at both these significance levels.

b) The individuals in the study are not random samples from low-fitness and high-fitness groups of middle-aged men. There are two ways in which they might be systematically different. The first is that all subjects in the study are college faculty. The "ego strength" personality factor for low-fitness and high-fitness middle-aged college faculty members might differ from the general population, and so might the mean difference between the groups. In addition, we don't have random samples of college faculty members. We are using volunteers in the study, and they might differ from both college faculty and the general population. It is hard to say in which direction this might bias the results, but the possibility of bias is definitely present.

Exercise 11.51

a) The hypotheses are $H_0: \mu_1 = \mu_2$ and $H_a: \mu_1 \neq \mu_2$, because the researchers were interested in a difference in either direction (Do VOT scores distinguish adults from children?). The standard error for the difference $\bar{x}_1 - \bar{x}_2$ between the mean VOT for children and adults is

$$\sqrt{\frac{s_1^2}{n_1} + \frac{s_2^2}{n_2}} = \sqrt{\frac{33.89^2}{10} + \frac{50.74^2}{20}} = 15.607.$$

Substituting this value into the denominator of the formula for t yields

$$t = \frac{\bar{x}_1 - \bar{x}_2}{\sqrt{\frac{s_1^2}{n_1} + \frac{s_2^2}{n_2}}} = \frac{-3.67 - (-23.17)}{15.607} = 1.25.$$

The smaller of $n_1 - 1 = 9$ and $n_2 - 1 = 19$ is 9 so we can refer the computed value of t to a $t(9)$ in Table C. The value 1.25 is between the critical values corresponding to upper tail probablities of 0.10 and 0.15. This gives a P-value between 0.20 and 0.30 because we need to double upper tail probabilities from the table on account of the test being two-sided. (The exact P-value for a t with 9 degrees of freedom using statistical software is 0.2428.) The data give no evidence of difference in mean VOT between children and adults.

b) The 95% confidence interval is $\bar{x}_1 - \bar{x}_2 \pm t^* \sqrt{\frac{s_1^2}{n_1} + \frac{s_2^2}{n_2}}$, in which t^* is the upper $(1-C)/2 = 0.025$ critical value for the t distribution with degrees of freedom equal to the smaller of $n_1 - 1 = 9$ and $n_2 - 1 = 19$. Using 9 degrees of freedom, we find that the value of $t^* = 2.262$ and the confidence interval is

$$-3.67 - (-23.17) \pm 2.262(15.607) = (-15.8, 54.8).$$

Because the interval includes the value 0, we fail to reject H_0 at the 5% level of significance. We knew this from part a because the P-value exceeded 5%.

SELECTED TEXT REVIEW EXERCISES

GUIDED SOLUTIONS

Exercise 11.57

KEY CONCEPTS: Matched pairs t test vs. two-sample t test, robustness of t procedures

a) Is the matched-pairs t test or two-sample t test appropriate here? Did the experimenter match the women in some way?

b) How many degrees of freedom are there? Use the smaller sample size minus one.

c) What's true about the sample sizes? What, if anything, can you say about outliers or skewness without seeing the original data?

Exercise 11.62

KEY CONCEPTS: One-sample t confidence intervals, checking assumptions

a) With a sample size of only n = 9, the most sensible graph would be a stemplot. Complete the stemplot below. Use split stems and just use the numbers to the left of the decimal place.

4 |
5 |
5 |
6 |
6 |

b) To compute a level C confidence interval, we use the formula $\bar{x} \pm t^* \frac{s}{\sqrt{n}}$, in which t^* is the upper $(1-C)/2$ critical value of the $t(n-1)$ distribution, which can be found in Table C. Fill in the missing values below. Don't forget to subtract one from the sample size when finding the appropriate degrees of freedom for the t confidence interval.

$C =$
$n =$
$t^* =$

Next, compute the values of $\bar{x}$ and s from the data given. Use statistical software or a calculator.

$\bar{x} =$ $s =$

Substitute all of these values into the formula to complete the computation of the 95% confidence interval.

$$\bar{x} \pm t^* \frac{s}{\sqrt{n}} =$$

COMPLETE SOLUTIONS

Exercise 11.57

a) The right test is the two-sample test. Each group, the control group and the undermine group, consists of 45 women. However, the 45 women in the two groups are different and there is no indication that the women in the two groups were matched in any way. Thus, there is no natural way to pair observations from the two groups together. Such pairing is necessary in a matched-pairs t test that uses the differences between matched pairs of observations to compute the t statistic.

b) Both groups have 45 observations apiece. If we use the conservative rule (degrees of freedom are one less than the smaller sample size), we would use $45 - 1 = 44$ degrees of freedom.

c) Each group consists of 45 observations. These are large samples (both are greater than 40), and so it is safe to use the t procedures even though the 7-point scale could not have a normal distribution. Note that with a 7-point scale (and sample means of around 4 or 5), the distribution of the observations could not have any extreme outliers and could not be strongly skewed. These are further reasons why the two sample t procedure should be reasonably accurate.

Exercise 11.62

a) The stemplot is given below. There are no outliers and the plot is skewed left. With this few observations, it is difficult to check the assumptions. In this case, we might still use the t procedures, but not with as much confidence in their validity as we had in most other examples.

```
4 | 9
5 | 1 4
5 |
6 | 0 3 3 4 4
6 | 5
```

b) An approximate 95% confidence interval for the mean percent of nitrogen in ancient air can be calculated from the data on the 9 specimens of amber. We use the formula for a t interval, namely $\bar{x} \pm t^* \frac{s}{\sqrt{n}}$. In this problem, $\bar{x}$ = 59.589, s = 6.2553, n = 9, hence t^* is the upper (1 – 0.95)/2 = 0.025 critical value for the $t(8)$ distribution. From Table C, we see that t^* = 2.306. Thus, the 95% confidence interval is

$$59.589 \pm 2.306 \frac{6.2553}{\sqrt{9}} = 59.589 \pm 4.808 = (54.78, 64.40)$$

Many statistical software packages and calculators will compute a confidence interval for you directly, after you input the data.

CHAPTER 12

INFERENCE FOR PROPORTIONS

SECTION 12.1

OVERVIEW

In this section, we consider inference about an unknown proportion p of a population that has some outcome. The outcome we are interested in is called a "success." The statistic that estimates the parameter p is the **sample proportion**

$$\hat{p} = \frac{\text{count of successes in the sample}}{\text{count of observations in the sample}}$$

obtained from an SRS of size n. As discussed in Section 9.2, for large samples, we can treat $\hat{p}$ as having a distribution that is approximately normal with mean $\mu = p$ and standard deviation $\sigma = \sqrt{p(1-p)/n}$.

In order to use the methods of this section for inference, the following assumptions need to be satisfied.

- The data are an SRS from the population of interest.

- The population is at least 10 times as large as the sample.

• For a test of H_0: $p = p_0$, the sample size n is so large that np_0 and $n(1 - p_0)$ are both 10 or more. For a confidence interval, n is so large that both the count of successes $n\hat{p}$ and the count of failures $n(1 - \hat{p})$ are 10 or more.

An **approximate level *C* confidence interval** for p is

$$\hat{p} \pm z^* \sqrt{\frac{\hat{p}(1-\hat{p})}{n}}$$

in which z^* is the upper (1–C)/2 critical value of the standard normal distribution,

$$\sqrt{\frac{\hat{p}(1-\hat{p})}{n}}$$

is the **standard error** of $\hat{p}$, and $z^* \sqrt{\frac{\hat{p}(1-\hat{p})}{n}}$ is the **margin of error**.

Tests of the hypothesis H_0: $p = p_0$ are based on the z **statistic**

$$z = \frac{\hat{p} - p_0}{\sqrt{\frac{p_0(1-p_0)}{n}}}$$

with P-values calculated from the $N(0, 1)$ distribution.

The **sample size** n required to obtain a confidence interval of approximate margin of error m for a proportion is

$$n = \left(\frac{z^*}{m}\right)^2 p^*(1 - p^*)$$

in which p^* is a guessed value for the population proportion and z^* is the upper (1 – C)/2 critical value of the standard normal distribution. To guarantee that the margin of error of the confidence interval is less than or equal to m no matter what the value of the population proportion may be, use a guessed value of $p^* = 1/2$, which yields

$$n = \left(\frac{z^*}{2m}\right)^2.$$

GUIDED SOLUTIONS

Exercise 12.3

KEY CONCEPTS: Parameters and statistics, proportions

a) To what group does the college president refer?

population = ______________________________________

parameter p = ______________________________________

b) A statistic is a number computed from a sample. What is the size of the sample, and how many in the sample support firing the coach? From these numbers compute

$\hat{p}$ =

Exercise 12.4

KEY CONCEPTS: When to use the procedures for inference about a proportion

Recall that the assumptions needed to safely use the methods of this section to compute a confidence interval are

- The data are an SRS from the population of interest.
- The population is at least 10 times as large as the sample.
- For a confidence interval, n is so large that both the count of successes $n\hat{p}$ and the count of failures $n(1 - \hat{p})$ are 10 or more.

These are the conditions that we must check in parts a through c. To do so, you will need to identify n and compute $\hat{p}$ in parts a through c.

a) n = $\qquad\qquad\qquad$ $\hat{p}$ =

Are the conditions met?

b) n = $\hat{p}$ =

Are the conditions met?

c) n = $\hat{p}$ =

Are the conditions met? Refer to Example 12.1 in the textbook for more about the National AIDS Behavioral Surveys.

Exercise 12.6

KEY CONCEPTS: Confidence intervals for a proportion

Recall that an approximate level C confidence interval for p, the true proportion of the month's orders that were shipped on time is

$$\hat{p} \pm z^* \sqrt{\frac{\hat{p}(1-\hat{p})}{n}}$$

in which z^* is the upper $(1 - C)/2$ critical value of the standard normal distribution.

a) From the information given in the problem, provide the values requested below.

n = sample size =

$\hat{p}$ = sample proportion of the month's orders shipped on time =

Can you safely use the methods of this section?

b) From the information given in the problem, provide the values requested below.

C = level of confidence requested =
z^* = upper (1 – C)/2 critical value of the standard normal distribution =

You will need to use Table A to find z^*. Now substitute these values into the formula for the confidence interval to complete the problem.

$$\hat{p} \pm z^* \sqrt{\frac{\hat{p}(1-\hat{p})}{n}} =$$

Exercise 12.11

KEY CONCEPTS: Sample size and margin of error

The sample size n required to obtain a confidence interval of approximate margin of error m for a proportion is

$$n = \left(\frac{z^*}{m}\right)^2 p^*(1-p^*)$$

in which p^* is a guessed value for the population proportion and z^* is the critical value of the standard normal distribution for the desired level of confidence. To apply this formula here, we must determine

m = desired margin of error =

p^* = a guessed value for the population proportion =

C = desired level of confidence =

z^* = the upper (1 – C)/2 critical value of the standard normal distribution

=

From the statement of the exercise, what are these values? Once you have determined them, use the formula to compute the required sample size n.

$$n = \left(\frac{z^*}{m}\right)^2 p^*(1-p^*) =$$

Exercise 12.15

KEY CONCEPTS: Testing hypotheses about a proportion

What statistical hypotheses should you test to answer this question?

H_0:
H_a:

Is the sample size n large enough so the test procedure can safely be applied?

$np_0 =$
$n(1 - p_0) =$

Compute the sample proportion of patients with "adverse symptoms,"

$\hat{p} =$

Next, compute the z test statistic

$$z = \frac{\hat{p} - p_0}{\sqrt{\frac{p_0(1-p_0)}{n}}} =$$

and the P-value of your test.

P-value =

What do you conclude?

Exercise 12.18

KEY CONCEPTS: Testing hypotheses about a proportion, confidence intervals for a proportion

a) First, state the hypotheses you will test in terms of p.

Next, compute the z test statistic (identify n and p_0 and calculate $\hat{p}$).

$$z = \frac{\hat{p} - p_0}{\sqrt{\frac{p_0(1-p_0)}{n}}} =$$

For your hypotheses,

P-value =

Is your result significant at the 5% level? (What must the P-value satisfy for this to be true?)

Finally, state your practical conclusions in the space below.

b) Recall that an approximate level C confidence interval for p is

$$\hat{p} \pm z^* \sqrt{\frac{\hat{p}(1-\hat{p})}{n}}$$

in which z^* is the upper $(1-C)/2$ critical value of the standard normal distribution. From the information given in the problem, provide the values requested below.

n = sample size =

$\hat{p}$ = sample proportion =

C = level of confidence requested =

z^* = upper $(1 - C)/2$ critical value of the standard normal distribution =

You will need to use Table A (or C) to find z^*. Next, substitute these values into the formula for the confidence interval to complete the problem.

$$\hat{p} \pm z^* \sqrt{\frac{\hat{p}(1-\hat{p})}{n}} =$$

c) If you are not sure how to answer this, review Section 2 of Chapter 5.

COMPLETE SOLUTIONS

Exercise 12.3

a) The population is presumably all 15,000 living alumni of the college. The parameter p is the proportion of these alumni who support firing the coach.

b) The statistic $\hat{p}$ is the proportion in the SRS that support firing the coach. It has value

$$\hat{p} = 76/200 = 0.38.$$

Exercise 12.4

a) $n = 50$ $\qquad\qquad$ $\hat{p} = 14/50 = 0.28.$

The data are an SRS from the population of interest.

The population consists of 175 students. This is *not* at least 10 times as large as the sample size $n = 50$.

The methods of this section *cannot* be safely used.

b) $n = 50$ $\qquad\qquad$ $\hat{p} = 38/50 = 0.76.$

The data are an SRS from the population of interest.

The population consists of 2400 students. This is at least 10 times as large as the sample size $n = 50$.

$n\hat{p} = 50(38/50) = 38$, $n(1-\hat{p}) = 50(1 - 38/50) = 50(12/50) = 12$. Both of these numbers are at least 10.

The methods of this section *can* be safely used.

c) $n = 2673$ $\qquad\qquad$ $\hat{p} = 0.002.$

The population consists of adult heterosexuals, a number in the millions. This is at least 10 times as large as the sample size $n = 2673$.

$n\hat{p}$ = 2673(0.002) = 5.346, $n(1-\hat{p})$ = 2673(1 – 0.002) = 2673(0.998) = 2667.654. The first of these numbers is less than 10.

The methods of this section *cannot* be safely used.

(Note: The actual sampling design was a complex stratified sample. The formulas for $\hat{p}$ are more involved than the one provided in this chapter. In this example, $n\hat{p}$ is not the number of "successes" which is why it is not an integer. The requirement that the sample be an SRS is not actually met either.)

Exercise 12.6

a) Because n = sample size = 100 and

$\hat{p}$ = sample proportion of the orders shipped on time = 86/100 = 0.86,

we have $n\hat{p} = 86$ and $n(1-\hat{p}) = 14$, so the methods in this section can safely be used.

b) C = level of confidence requested = 0.95

z^* = upper (1–C)/2 critical value of the standard normal distribution

= upper 0.025 critical value of the standard normal distribution = 1.96

Substituting these values into the formula for the confidence interval yields

$$\hat{p} \pm z^* \sqrt{\frac{\hat{p}(1-\hat{p})}{n}} \qquad = 0.86 \pm (1.96)\sqrt{\frac{(0.86)(1-0.86)}{100}}$$

$$= 0.86 \pm (1.96)(0.0347) = 0.86 \pm 0.068$$

or 0.792 to 0.928.

Exercise 12.11

We start with the guess that $p^* = 0.75$. For 95% confidence we use $z^* = 1.96$. The sample size we need for a margin of error $m = 0.04$ is thus

$$n = \left(\frac{z^*}{m}\right)^2 p^*(1-p^*) = \left(\frac{1.96}{0.04}\right)^2 0.75(1-0.75) = 450.1875.$$

We round this up to $n = 451$. Thus, a sample of size 451 is needed to estimate the proportion of Americans with at least one Italian grandparent who can taste PTC, to within ±.04 with 95% confidence.

Exercise 12.15

The question being asked can be restated as one in which we wish to test the hypotheses

$$H_0: p = 0.10, \;\; H_a: p < 0.10.$$

Since $np_0 = 440(0.10) = 44$ and $n(1 - p_0) = 440(1 - 0.1) = 396$ are both 10 or more the z test procedure can safely be used.

The sample proportion experiencing adverse symptoms is

$$\hat{p} = 23/440 = 0.05227.$$

The z test statistic is therefore

$$z = \frac{\hat{p} - p_0}{\sqrt{\dfrac{p_0(1-p_0)}{n}}} = \frac{0.05227 - 0.10}{\sqrt{\dfrac{0.10(1-0.10)}{440}}} = \frac{-0.04773}{.0143} = -3.34.$$

The P-value is the area to the left of –3.34 in the z table. Going to the table, we find that the P-value = 0.0004. We would conclude that there is strong statistical evidence that the proportion of patients suffering adverse symptoms is less than 10%.

Exercise 12.18

a) We wish to see whether the majority (more than half) of people prefer the taste of fresh-brewed coffee. The hypotheses to be tested are therefore

$$H_0: p = 0.50, \;\; H_a: p > 0.50.$$

We see that $n = 50$, $p_0 = 0.50$, $\hat{p} = 31/50 = 0.62$, so the z test statistic is

$$z = \frac{\hat{p} - p_0}{\sqrt{\dfrac{p_0(1-p_0)}{n}}} = \frac{0.62 - 0.50}{\sqrt{\dfrac{0.50(1-0.50)}{50}}} = \frac{0.12}{\sqrt{0.005}} = 1.70.$$

The P-value for this value of the z test statistic is the area to the right of 1.70 under a standard normal curve. From Table A, this is

$$P\text{-value} = 0.0446.$$

Because this is smaller than 0.05, the result is significant at the 5% level.

We may conclude that there is good statistical evidence that the majority (more than half) of people prefer the taste of fresh-brewed coffee. This conclusion is based on data from 50 subjects of which 62% favored the taste of fresh-brewed coffee.

b) From the information given in the problem,

n = sample size = 50

$\hat{p}$ = sample proportion = 0.62

C = level of confidence requested = 0.90

z^* = upper $(1 - C)/2$ critical value of the standard normal distribution = 1.645

Thus the 90% confidence interval for p is

$$\hat{p} \pm z^* \sqrt{\frac{\hat{p}(1-\hat{p})}{n}} = 0.62 \pm 1.645\sqrt{\frac{0.62(1-0.62)}{50}}$$

$$= 0.62 \pm 1.645\sqrt{0.004712}$$

$$= 0.62 \pm 0.113$$

c) You should present the two cups of coffee to the subjects in a random order (i.e., determine which cup a subject gets first by a random mechanism, such as flipping a coin).

SECTION 12.2

OVERVIEW

Confidence intervals and tests designed to compare two population proportions are based on the **difference in the sample proportions** $\hat{p}_1 - \hat{p}_2$. The formula for the level C confidence interval is

$$(\hat{p}_1 - \hat{p}_2) \pm z^*\text{SE}$$

in which z^* is the upper $(1 - C)/2$ standard normal critical value and SE is the standard error for the difference in the two proportions computed as

$$\text{SE} = \sqrt{\frac{\hat{p}_1(1-\hat{p}_1)}{n_1} + \frac{\hat{p}_2(1-\hat{p}_2)}{n_2}}$$

In practice, use this confidence interval when the populations are at least 10 times as large as the samples and the counts of successes and failures are 5 or more in both samples.

Significance tests for the equality of the two proportions, H_0: $p_1 = p_2$, use a different standard error for the difference in the sample proportions, which is based on a **pooled estimate** of the common (under H_0) value of p_1 and p_2,

$$\hat{p} = \frac{\text{count of successes in both samples combined}}{\text{count of observations in both samples combined}}.$$

The test uses the z statistic

$$z = \frac{\hat{p}_1 - \hat{p}_2}{\sqrt{\hat{p}(1-\hat{p})\left(\frac{1}{n_1} + \frac{1}{n_2}\right)}}$$

and P-values are computed using Table A of the standard normal distribution. In practice, use this test when the populations are at least 10 times as large as the samples and the counts of successes and failures are 5 or more in both samples.

GUIDED SOLUTIONS

Exercise 12.21

KEY CONCEPTS: Confidence intervals for the difference between two population proportions

a) Let p_1 represent the proportion of white Protestants who believed free speech included the right to make speeches in favor of communism, and p_2 the proportion of white Catholics who believed this. Recall that a level C confidence interval for $p_1 - p_2$ is

$$(\hat{p}_1 - \hat{p}_2) \pm z^*\text{SE}$$

in which z^* is the upper $(1 - C)/2$ standard normal critical value and SE is the standard error for the difference in the two proportions computed as

$$SE = \sqrt{\frac{\hat{p}_1(1-\hat{p}_1)}{n_1} + \frac{\hat{p}_2(1-\hat{p}_2)}{n_2}}$$

We guide you through the steps needed to compute such a confidence interval.

The two sample sizes are

$n_1 =$

$n_2 =$

From the data, the estimates of these two proportions are (complete the following)

$\hat{p}_1 =$

$\hat{p}_2 =$

Next, compute the standard error

$$SE = \sqrt{\frac{\hat{p}_1(1-\hat{p}_1)}{n_1} + \frac{\hat{p}_2(1-\hat{p}_2)}{n_2}} =$$

For a 95% confidence interval

$z^* =$

Next, compute the interval

$(\hat{p}_1 - \hat{p}_2) \pm z^*SE =$

b) Check that it is safe to use the z confidence interval here. (Under what conditions is the z confidence interval applicable?)

Exercise 12.26

KEY CONCEPTS: Testing equality of two population proportions, confidence intervals for the difference between two populations proportions

a) Let p_1 represent the proportion of students from urban/suburban backgrounds who succeed and p_2 represent the proportion from rural/small-town backgrounds who succeed. Recall that a test of the hypothesis H_0: $p_1 = p_2$, uses the z statistic

$$z = \frac{\hat{p}_1 - \hat{p}_2}{\sqrt{\hat{p}(1-\hat{p})\left(\frac{1}{n_1} + \frac{1}{n_2}\right)}}$$

in which n_1 and n_2 are the sizes of the samples, $\hat{p}_1$ and $\hat{p}_2$ the estimates of p_1 and p_2, and

$$\hat{p} = \frac{\text{count of successes in both samples combined}}{\text{count of observations in both samples combined}}$$

We guide you through the steps needed to carry out the test.

First, state the hypotheses to be tested. (What is the alternative in this case: one-sided or two-sided?)

The two sample sizes are

$n_1 =$

$n_2 =$

From the data, the estimates of these two proportions are (complete the following)

$\hat{p}_1 =$

$\hat{p}_2 =$

Next, compute

$$\hat{p} = \frac{\text{count of successes in both samples combined}}{\text{count of observations in both samples combined}} =$$

and then the value of the z test statistic

$$z = \frac{\hat{p}_1 - \hat{p}_2}{\sqrt{\hat{p}(1-\hat{p})\left(\frac{1}{n_1} + \frac{1}{n_2}\right)}} =$$

Finally, compute the P-value using Table A

P-value =

What do you conclude?

b) Construct the 90% confidence interval, using the steps outlined below to guide you.

The two sample sizes are

$n_1 =$

$n_2 =$

From the data, the estimates of these two proportions are (complete the following)

$\hat{p}_1 =$

$\hat{p}_2 =$

Next, compute the standard error

$$\text{SE} = \sqrt{\frac{\hat{p}_1(1-\hat{p}_1)}{n_1} + \frac{\hat{p}_2(1-\hat{p}_2)}{n_2}} =$$

For a 90% confidence interval

$$z^* =$$

Now compute the interval

$$(\hat{p}_1 - \hat{p}_2) \pm z^*\text{SE} =$$

Exercise 12.28

KEY CONCEPTS: Testing equality of two population proportions

We guide you through the steps needed to carry out the test.

First, state the hypotheses to be tested. (What is the alternative in this case: one-sided or two-sided?)

The two sample sizes are

$$n_1 =$$

$$n_2 =$$

From the data, the estimates of these two proportions are (complete the following)

$$\hat{p}_1 =$$

$$\hat{p}_2 =$$

Next, compute

$$\hat{p} = \frac{\text{count of successes in both samples combined}}{\text{count of observations in both samples combined}} =$$

and then the value of the z test statistic

$$z = \frac{\hat{p}_1 - \hat{p}_2}{\sqrt{\hat{p}(1-\hat{p})\left(\frac{1}{n_1} + \frac{1}{n_2}\right)}} =$$

Finally, compute the *P*-value using Table A

P-value =

What do you conclude?

COMPLETE SOLUTIONS

Exercise 12.21

a) The two sample sizes are the number of Protestants and Catholics, and these are

$$n_1 = 267$$

$$n_2 = 230$$

From the data, the estimates of these two proportions are (complete the following)

$$\hat{p}_1 = 104/267 = 0.390$$

$$\hat{p}_2 = 75/230 = 0.326$$

The standard error is

$$\text{SE} = \sqrt{\frac{\hat{p}_1(1-\hat{p}_1)}{n_1} + \frac{\hat{p}_2(1-\hat{p}_2)}{n_2}} = \sqrt{\frac{0.390(1-0.390)}{267} + \frac{0.326(1-0.326)}{230}}$$

$$= \sqrt{0.000891 + 0.000955} = 0.043$$

and for a 95% confidence interval

$$z^* = 1.96$$

so our 95% confidence interval is

$$(\hat{p}_1 - \hat{p}_2) \pm z^*SE = (0.390 - 0.326) \pm 1.96(0.043) = 0.064 \pm 0.084.$$

$$= (-0.020, 0.148).$$

b) We can use z methods here because the population of all white Protestants and Catholics in Detroit is much larger than (at least ten times as large) as the samples obtained, and because the counts of successes and failures are 104 and 163 for sample 1 (both larger than 5) and 75 and 155 for sample 2 (both larger than 5).

Exercise 12.26

a) We are interested in determining whether there is good evidence that the proportion of students who succeed is *different* for urban/suburban versus rural/small-town backgrounds. Thus the hypotheses to be tested are

$$H_0: p_1 = p_2$$

$$H_a: p_1 \neq p_2.$$

The two sample sizes are

$$n_1 = \text{number from urban/suburban background} = 65$$

$$n_2 = \text{number from rural/small-town background} = 55.$$

From the data, the estimates of the two proportions of students who succeeded are

$$\hat{p}_1 = 52/65 = 0.8$$

$$\hat{p}_2 = 30/55 = 0.545.$$

Next, we compute

$$\hat{p} = \frac{\text{count of successes in both samples combined}}{\text{count of observations in both samples combined}} = \frac{52+30}{65+55}$$

$$= 82/120 = 0.683.$$

The value of the z test statistic

$$z = \frac{\hat{p}_1 - \hat{p}_2}{\sqrt{\hat{p}(1-\hat{p})\left(\frac{1}{n_1} + \frac{1}{n_2}\right)}} = \frac{0.8 - 0.545}{\sqrt{0.683(1-0.683)\left(\frac{1}{65} + \frac{1}{55}\right)}}$$

$$= \frac{0.255}{\sqrt{0.00727}} = 2.99.$$

Using Table A (we need to double the tail area because this is a two-sided test)

$$P\text{-value} = 2 \times (0.0014) = 0.0028.$$

There is good statistical evidence that the proportion of students who succeed is different for urban/suburban versus rural/small-town backgrounds.

b) The two sample sizes are as in part a, namely

$$n_1 = 65$$

$$n_2 = 55.$$

The estimates of the two proportions are also as in part a, namely

$$\hat{p}_1 = 52/65 = 0.8$$

$$\hat{p}_2 = 30/55 = 0.545.$$

The standard error is

$$\text{SE} = \sqrt{\frac{\hat{p}_1(1-\hat{p}_1)}{n_1} + \frac{\hat{p}_2(1-\hat{p}_2)}{n_2}} = \sqrt{\frac{0.8(1-0.8)}{65} + \frac{0.545(1-0.545)}{55}}$$

$$= \sqrt{0.00246 + 0.00451} = 0.08$$

For a 90% confidence interval

$$z^* = 1.645$$

so our confidence interval is

$$(\hat{p}_1 - \hat{p}_2) \pm z^*\text{SE} = (0.8 - 0.545) \pm 1.645(0.08) = 0.255 \pm 0.13.$$

Exercise 12.28

Let p_1 represent the proportion of female students who will succeed and p_2 the proportion of males who will succeed. We are interested in determining whether there is a difference in these two proportions, hence we should test the hypotheses

$$H_0\text{: } p_1 = p_2$$

$$H_a\text{: } p_1 \neq p_2.$$

The two sample sizes are

n_1 = number of female students in the course = 34

n_2 = number of male students in the course = 89.

From the data, the estimates of these two proportions are

$$\hat{p}_1 = 23/34 = 0.6765$$

$$\hat{p}_2 = 60/89 = 0.6742.$$

Next, we compute

$$\hat{p} = \frac{\text{count of successes in both samples combined}}{\text{count of observations in both samples combined}} = \frac{23+60}{34+89}$$

$$= 83/123 = 0.6748.$$

Thus, the value of the z test statistic is

$$z = \frac{\hat{p}_1 - \hat{p}_2}{\sqrt{\hat{p}(1-\hat{p})\left(\frac{1}{n_1}+\frac{1}{n_2}\right)}} = \frac{0.6765 - 0.6742}{\sqrt{0.6748(1-0.6748)\left(\frac{1}{34}+\frac{1}{89}\right)}}$$

$$= \frac{0.0023}{\sqrt{0.00892}} = 0.02.$$

Finally, we compute the P-value using Table A (we need to double the tail area because this is a two-sided test)

$$P\text{-value} = 2 \times (0.4920) = 0.9840$$

The data provide no real statistical evidence of a difference between the proportion of men and women who succeed.

SELECTED TEXT REVIEW EXERCISES

GUIDED SOLUTIONS

Exercise 12.35

KEY CONCEPTS: Interpreting confidence intervals

What are the assumptions for inference about a proportion? Review the list given in Section 1 of the text. Are any of these assumptions violated?

COMPLETE SOLUTIONS

Exercise 12.35

In order for the procedures of this chapter to be safe, samples should be simple random samples from the populations of interest. Call-in polls are not simple random samples. You will recall from Chapter 5 that call-in surveys are often biased. Both the value of 81% and the calculations that produce the confidence interval should not be trusted. Fancy statistical procedures do not fix poorly designed surveys.

CHAPTER 13

INFERENCE FOR TABLES: CHI-SQUARE PROCEDURES

SECTION 13.1

OVERVIEW

The **chi-square goodness of fit** can determine whether a population has a certain hypothesized distribution. The hypothesized distribution can be expressed as percents of the population falling into each of several outcome categories. The hypotheses tested by the chi-square goodness of fit are

H_0: the actual population percents are *equal* to the hypothesized percentages

H_a: the actual population percents are *different from* the hypothesized percentages.

The observed sample counts (O) of each of the variable outcome categories are compared with the **expected counts (E)** computed under the hypothesized distribution. The expected count for any variable category is obtained by multiplying the percent of the reference distribution for each category by the sample size. The test procedure is based on the **chi-square (χ^2) statistic**,

$$\chi^2 = \sum \frac{(O - E)^2}{E},$$

in which the sum is over the n variable categories. The value of the χ^2 statistic is compared with critical values from the **chi-square distribution** with $n - 1$ **degrees of freedom.** If the result is statistically signficant, then do a follow-up analysis that compares the observed counts with the expected counts and looks for the largest **components of chi-square.** You may safely use the chi-square distribution as an approximation to the distribution of the statistic χ^2 when all individual expected cell counts are at least 1 and no more than 20% of the expected counts are less than 5.

GUIDED SOLUTIONS

Exercise 13.3

KEY CONCEPTS: Test for goodness of fit

a) What is the hypothesized distribution of marital status? Use it to state the null and alternative hypotheses in words.

H_0:

H_a:

b) Find the expected counts by completing the table below. Remember that the expected count for any outcome is the hypothesized percent times the sample size. What is the sample size in this example?

Marital status	1996 Percents	Expected counts
Never married		
Married		
Widowed		
Divorced		

c) Calculate the goodness of fit statistic by completing the table below. The observed counts are obtained from the sample, and you can use the expected counts computed in part b. What are the degrees of freedom for the χ^2 statistic in this case?

Marital status	Observed O	Expected E	$(O-E)^2/E$
Never married			
Married			
Widowed			
Divorced			

Adding the last column we get χ^2 = ____________.

d) The P-value is the area to the right of the value of χ^2 computed in part c using the line in Table E corresponding to the degrees of freedom found in part c.

COMPLETE SOLUTIONS

Exercise 13.3

a) The hypothesized distribution is the distribution of marital status of the U.S. adult population in 1996.

H_0: The marital status distribution of 25- to 29-year-old males in the U.S. is the *same* as the marital status distribution of the U.S. adult population in 1996.

H_a: The marital status distribution of 25- to 29-year-old males in the U.S. is *different from* the marital status distribution of the U.S. adult population in 1996.

b) The sample size is 500. The 1996 percents are the hypothesized percents and are given in the Table in the textbook. The column of expected counts is obtained by multiplying each percent by the sample size of 500.

Marital status	1996 Percents	Expected counts
Never married	23.26	116.30
Married	60.31	301.55
Widowed	7.00	35.00
Divorced	9.43	47.15

c) There are $n = 4$ variable categories, so the degrees of freedom is $n - 1 = 3$.

Marital status	Observed O	Expected E	$(O-E)^2/E$
Never married	260	116.30	177.555
Married	220	301.55	22.054
Widowed	0	35.00	35.000
Divorced	20	47.15	15.633

Adding the last column, we get $\chi^2 = 177.555 + 22.054 + 35.000 + 15.633$
$= 250.24$.

d) From Table E, the area to the right of 17.73 is 0.0005. Because 250.24 is larger than 17.73, we can say that the P-value is less than 0.0005. Using software or a calculator, we have the P-value = 0.0000. There is very strong evidence that the marital status distribution of 25- to 29-year-old males in the U.S. is *different from* the marital status distribution of the U.S. adult population in 1996.

SECTION 13.2

OVERVIEW

The inference methods of Chapter 12 are extended to a comparison of more than two population proportions. When comparing more than two proportions, it is best to first do an overall test to see whether there is good evidence of any differences among the proportions. Then a detailed follow-up analysis can be performed to decide which proportions are different and to estimate the sizes of the differences.

The overall test for comparing several population proportions arranges the data in a **two-way table**. Two-way tables were introduced in Chapter 4; the tables are a way of displaying the relationship between any two categorical variables. The tables are also called ***r* × *c* tables**, in which *r* is the number of rows and *c* is the number of columns. Although it is not necessary, often the rows of the two-way table correspond to the populations or treatment groups, and the columns to different categories of the response. When comparing several proportions with this assignment of rows and columns, there would be only two columns representing the counts of successes and failures, the two response categories.

The null hypothesis is $H_0: p_1 = p_2 = \ldots = p_n$, which says that the n population proportions are the same. The alternative is "many sided," as the proportions can differ from each other in a variety of ways. To test this hypothesis, we will compare the **observed counts** with the **expected counts** when H_0 is true. The expected cell counts are computed using the formula

$$\text{Expected count} = \frac{\text{row total} \times \text{column total}}{n}$$

in which n is the total number of observations.

The statistic we will use to compare the expected counts with the observed counts is the **chi-square statistic**. It measures how far the observed and expected counts are from each other using the formula

$$\chi^2 = \sum \frac{(\text{observed} - \text{expected})^2}{\text{expected}}$$

in which we sum up all the $r \times c$ cells in the table.

When the null hypothesis is true, the distribution of the test statistic χ^2 is approximately the chi-squared distribution with $(r - 1)(c - 1)$ degrees of freedom. The P-value is the area to the right of χ^2 under the chi-square density curve. Use Table E in the back of the book to get the critical values and to compute the *P*-value.

We can use the chi-squared statistic when the data satisfy the following conditions.

- The data are independent SRSs from several populations, and each observation is classified according to one categorical variable.
- The data are from a single SRS and each observation is classified according to two categorical variables.
- No more than 20% of the cells in the two-way table have expected counts less than 5.
- All cells have an expected count of at least 1.
- In the special case of the 2×2 table, all expected counts should exceed 5.

GUIDED SOLUTIONS

Exercise 13.13

KEY CONCEPTS: $r \times c$ tables, expected counts, comparison of more than two proportions.

a) What are the values of *r* and *c*?

$r =$ $c =$

b) To compute the proportion of successful students in each of the three extracurricular activity groups, it is easiest to first fill in the column totals in the table. Note that in this example, the columns correspond to the "populations" and the rows to successes or failures.

	< 2	2 to 12	> 12
C or better	11	68	3
D or F	9	23	5
Totals			

Next, for each of the three extracurricular activity groups, compute the proportion of successful students (C or better).

Activity group	Proportion of successful students
<2 hours	
2 to 12 hours	
>12 hours	

What kind of relationship between extracurricular activities and succeeding in the course is shown by these proportions?

c) Complete the bar chart to compare the three proportions of successes found in part b.

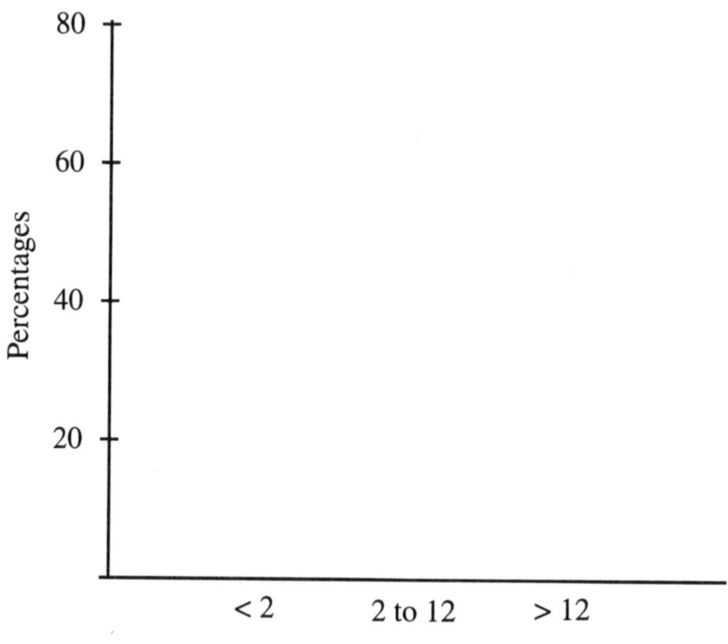

d) Fill in the expected counts in the following table.

	< 2	2 to 12	> 12
C or better	13.78		
D or F			

Remember that the expected cell counts are computed using the formula

$$\text{Expected count} = \frac{\text{row total} \times \text{column total}}{n}$$

in which n is the total number of observations. The value in the upper left-hand corner was obtained as

$$\text{Expected count} = \frac{\text{row total} \times \text{column total}}{n} = \frac{(82)(20)}{119} = 13.78$$

e) Next, compare the observed and expected counts. Are any deviations unusually large? Do the deviations follow the pattern found in part b?

Exercise 13.15

KEY CONCEPTS: Computing the χ^2 statistic

a) We reproduce the information from the Minitab output from the text below. Expected counts are printed below observed counts.

	<2	2 to 12	>12	Total
A, B, C	11	68	3	82
	13.78	62.71	5.51	
D or F	9	23	5	37
	6.22	28.29	2.49	
Total	20	91	8	119

Chi-Sq = 0.561 + 0.447 + 1.145 + 1.244 + 0.991 + 2.538 = 6.926
df = 2 1 cells with expected counts less than 5.0

Chisquare 2.
6.9260 0.9687

The expected counts were verified in exercise 13.13. The components of the chi-square statistic are calculated from the expected and observed count in each cell. If you use the expected counts in the table to compute the components of the chi-square statistic, your results will be quite close but not exactly equal to the values in the Minitab output. This is because the printed output rounds off the expected counts to two decimal places, but the calculations used by Minitab to compute the components of the chi-square statistic are based on "unrounded" values. So if your calculations are based on the "rounded off" values, they will not agree exactly with the values in the Minitab output. Verify the components of the chi-square statistic. We have verified the entry for row 1 and column 1.

Cell in row 1, column 1: $\dfrac{(\text{observed count} - \text{expected count})^2}{\text{expected count}} =$

$$\frac{(11 - 13.78)^2}{13.78} = 0.561$$

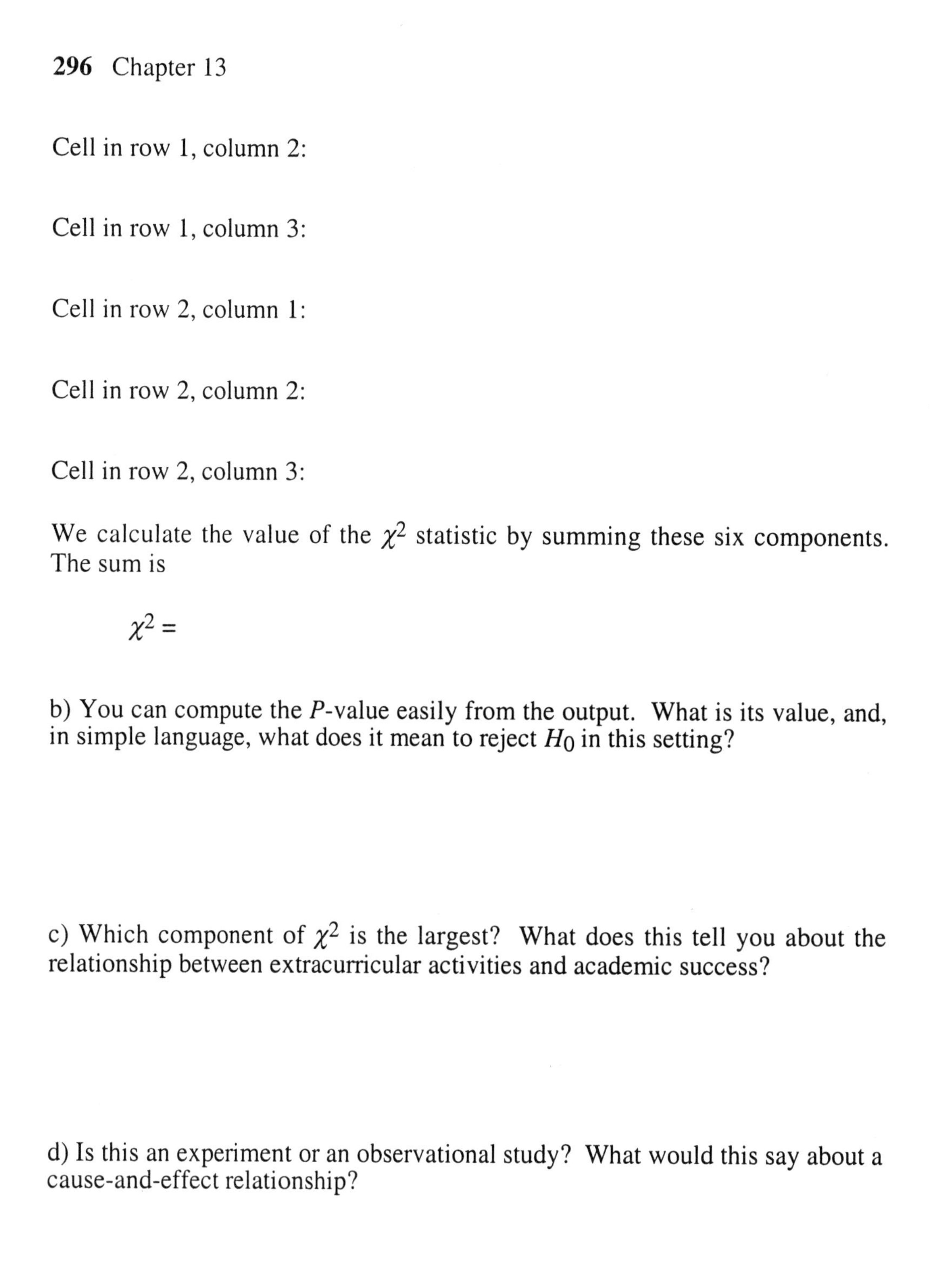

Cell in row 1, column 2:

Cell in row 1, column 3:

Cell in row 2, column 1:

Cell in row 2, column 2:

Cell in row 2, column 3:

We calculate the value of the χ^2 statistic by summing these six components. The sum is

$$\chi^2 =$$

b) You can compute the *P*-value easily from the output. What is its value, and, in simple language, what does it mean to reject H_0 in this setting?

c) Which component of χ^2 is the largest? What does this tell you about the relationship between extracurricular activities and academic success?

d) Is this an experiment or an observational study? What would this say about a cause-and-effect relationship?

Exercise 13.17

KEY CONCEPTS: Degrees of freedom for the chi-square distribution, *P*-values using Table E, mean of the chi-square distribution

a) What are the values of r and c in the table? The degrees of freedom can be found using the formula

$$\text{Degrees of freedom} = (r - 1)(c - 1) =$$

b) Go to the row in Table E corresponding to the degrees of freedom found in part a. Between what two entries in Table E does the value $\chi^2 = 6.926$ lie? What does this tell you about the *P*-value?

COMPLETE SOLUTIONS

Exercise 13.13

a) There are $r = 2$ rows ("C or better" and "D or F") and $c = 3$ columns ("<2," "2 to 12," and ">12").

b) If we add the counts (students) in each of the three groups, we find that

	< 2	2 to 12	> 12
C or better	11	68	3
D or F	9	23	5
Totals	20	91	8

From the group (column) totals, we calculate the proportion of successful students in each column.

	< 2	2 to 12	> 12
C or better	11/20 = .550	68/91 = .747	3/8 = .375

The proportions seem to suggest that spending too much time in extracurricular activities is associated with a lack of success in the course. Spending little time in extracurricular activities is more highly associated with success, while spending a moderate amount of time is the group most highly associated with success. Perhaps the maxim "Moderation in all things" applies here!

c) Here is the bar chart of the percentages in part b.

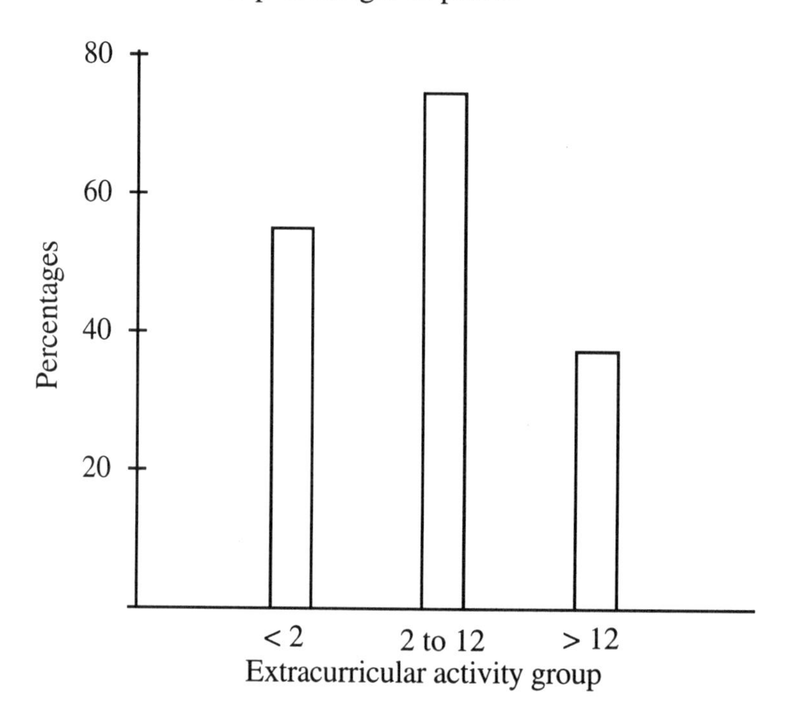

d) Following is the table of expected counts, with the required calculations.

	< 2	2 to 12	> 12
C or better	13.78	62.71	5.51
D or F	6.22	28.29	2.49

$62.71 = \frac{(82)(91)}{119}$ $5.51 = \frac{(82)(8)}{119}$ $6.22 = \frac{(37)(20)}{119}$

$28.29 = \frac{(37)(91)}{119}$ $2.49 = \frac{(37)(8)}{119}$

e) Here is a table presenting the observed and expected counts together for successful students.

	< 2	2 to 12	> 12
C or better	Observed = 11 Expected = 13.78	Observed = 68 Expected = 62.71	Observed = 3 Expected = 5.51
D or F	Observed = 9 Expected = 6.22	Observed = 23 Expected = 28.29	Observed = 5 Expected = 2.49

None of the deviations is enormous, but neither is there extremely close agreement between the observed and expected counts. The largest deviations between the observed and expected counts (in an absolute sense) occur for the 2 to 12 hours in the extracurricular activities group. The observed number of successful students is larger than expected and the observed number of unsuccessful students is smaller than expected. This may suggest an association between success in schoolwork and a moderate involvement in extracurricular activities, as was found in part b.

Exercise 13.15

a) The components of the chi-square statistic are calculated from the expected and observed count in each cell.

Cell in row 1, column 1: $\dfrac{(\text{observed count} - \text{expected count})^2}{\text{expected count}} = \dfrac{(11 - 13.78)^2}{13.78}$

$= 0.561$

Cell in row 1, column 2: $\dfrac{(\text{observed count} - \text{expected count})^2}{\text{expected count}} = \dfrac{(68 - 62.71)^2}{62.71}$

$= 0.446$

Cell in row 1, column 3: $\dfrac{(\text{observed count} - \text{expected count})^2}{\text{expected count}} = \dfrac{(3 - 5.51)^2}{5.51}$

$= 1.143$

Cell in row 2, column 1: $\dfrac{(\text{observed count} - \text{expected count})^2}{\text{expected count}} = \dfrac{(9 - 6.22)^2}{6.22}$

$= 1.243$

Cell in row 2, column 2: $\dfrac{(\text{observed count} - \text{expected count})^2}{\text{expected count}} = \dfrac{(23 - 28.29)^2}{28.29}$

$= 0.989$

Cell in row 2, column 3: $\dfrac{(\text{observed count} - \text{expected count})^2}{\text{expected count}} = \dfrac{(5 - 2.49)^2}{2.49}$

$= 2.530$

We calculate the value of the χ^2 statistic by summing these six components. The sum is

$$\chi^2 = 6.912.$$

b) The *P*-value is the probability of obtaining a value of the χ^2 statistic at least as large as the observed value (6.926 on the Minitab output). The probability of a chi-square value with 2 df being less than 6.926 is given on the output as 0.9687. Subtracting from 1 gives the *P*-value = 0.0313. We would reject H_0 at level 0.05 but not at level 0.01. Rejecting H_0 means that the observed differences in the proportions in each group for the successful ("C or better") and unsuccessful students ("D or F") cannot be easily attributed to chance. In other words, there is evidence that the proportions of successful students in the three groups (representing amounts of time spent on extracurricular activities) are different.

c) The term contributing most to χ^2 is that of the number of unsuccessful students spending more than 12 hours per week on extracurricular activities. This points to the fact that if one spends too much time on extracurricular activities, one has little time for schoolwork and is likely to be unsuccessful.

d) The study was not a designed experiment and hence does not prove that spending more or less time on extracurricular activity *causes* changes in academic success. It is an observational study. While at first glance it seems plausible that changing the amount of time spent on extracurricular activities ought to cause changes in academic success, this assumes (among other things) that students will trade time spent on schoolwork with time spent on extracurricular activities. However, some students may seek out involvement in extracurricular activities as an escape from schoolwork. If they aren't involved in extracurricular activities, they will simply "goof off." Thus, changes in extracurricular activity will not necessarily produce changes in academic performance.

Exercise 13.17

a) As we saw in exercise 13.13, there are $r = 2$ rows ("C or better" and "D or F") and $c = 3$ columns ("<2," "2 to 12," and ">12"). Thus, the number of degrees of freedom is

$$(r-1)(c-1) = (2-1)(3-1) = (1)(2) = 2$$

which agrees with the value in the Minitab output (see exercise 13.15).

b) If we look in the df = 2 row of Table E, we find the following information.

p	.05	.025
x^*	5.99	7.38

$\chi^2 = 6.926$ lies between the entries for $p = .05$ and $p = .025$. This tells us that the *P*-value is between .05 and .025.

SELECTED TEXT REVIEW EXERCISES

GUIDED SOLUTIONS

Exercise 13.37

a) Review the methods of Section 12.1 for finding a confidence interval for a proportion. The formula for an approximate level C confidence interval for p is

$$\hat{p} \pm z^* \sqrt{\frac{\hat{p}(1-\hat{p})}{n}} =$$

in which $\hat{p}$ is the sample proportion who thought that the military should have more control over how news organizations report about the war and z^* is the upper $(1 - C)/2$ critical value of the standard normal distribution.

b) This is a test of the equality of *two* proportions: the proportion of adults who thought the military should have more control and the proportion of students at the University of Virginia who felt this way. This can be done using the chi-square procedures of this chapter viewing the data as a 2×2 table, or using the z procedure of Section 12.2 for testing equality of two proportions. In either case, the formulas for the procedures are expressed most easily in terms of the counts of successes, not the proportion of successes. Determining the counts from the percents and the sample sizes is the first step in either analysis. How can you determine the count or number in the sample who thought the military should have more control from the corresponding percent and sample size?

Population	Sample size	Percent	Count
Adults (1991)	924	57%	
Students at UV	174	55%	

In this part, we illustrate the use of χ^2 to test the equality of two proportions, and in part c we illustrate the use of the z procedure.

Complete the 2×2 table given below. Include the counts as well as the expected counts in each cell. We have provided the counts for the first row and the expected count for the first cell.

	Yes	No	Total
Adults (1991)	527 524.27	397	924
Students at UV			
Total			

Next,

$$\chi^2 = \sum \frac{(\text{observed} - \text{expected})^2}{\text{expected}} =$$

in which you need to remember to sum over all 4 cells in the table. What do you conclude?

c) In this part, we illustrate the use of the z procedure of Section 12.2 for testing equality of two proportions who felt the military should have more control using the two different wordings for the question. Determining the counts from the percents and the sample sizes is the first step.

Population	Sample size	Percent	Count
Student group 1	174	55%	
Student group 2	199	16%	

From the data, the estimates of these two proportions are (complete the following)

$\hat{p}_1 =$

$\hat{p}_2 =$

Next, compute

$$\hat{p} = \frac{\text{count of successes in both samples combined}}{\text{count of observations in both samples combined}} =$$

and the value of the z test statistic

$$z = \frac{\hat{p}_1 - \hat{p}_2}{\sqrt{\hat{p}(1-\hat{p})\left(\frac{1}{n_1} + \frac{1}{n_2}\right)}} =$$

d) Testing equality of three proportions requires the chi-square procedures of this chapter. First, organize the data into a 3 × 2 table, giving the counts that said yes and no for each of the three groups. Give the expected counts below the counts for each cell as well.

	Yes	No	Total
Adults (1991)			
Student Group 1			
Student Group 2			
Total			

Next,

$$\chi^2 = \sum \frac{(\text{observed} - \text{expected})^2}{\text{expected}} =$$

in which you need to remember to sum over all six cells in the table. What do you conclude?

COMPLETE SOLUTIONS

Exercise 13.37

a) For a 90% confidence interval, $C = 0.90$ and $z^* = 1.645$ is the upper $(1 - C)/2 = 0.05$ critical value of the standard normal distribution. The sample proportion is given as $\hat{p} = 0.57$. Inserting these values into the formula for the confidence interval yields

$$\hat{p} \pm z^* \sqrt{\frac{\hat{p}(1-\hat{p})}{n}} = 0.57 \pm 1.645\sqrt{\frac{(0.57)(0.43)}{924}} = 0.57 \pm .028 = (0.542, 0.598).$$

b) Because the percent is $\hat{p}$, and

$$\hat{p} = \frac{\text{count of successes in the sample}}{\text{count of observations in the sample}}$$

the count of successes in the sample is obtained by multiplying the percent by the sample size. For example, for adults in 1991 the count is $924(0.57) = 527$.

Population	Sample size	Percent	Count
Adults (1991)	924	57%	527
Students at UV	174	55%	96

The 2 × 2 table is

	Yes	No	Total
Adults (1991)	527 524.27	397 399.73	924
Students at UV	96 98.73	78 75.27	174
Total	623	475	1098

The expected values are obtained as

$$524.27 = \frac{(924)(623)}{1098} \qquad 399.73 = \frac{(924)(475)}{1098}$$

$$98.73 = \frac{(174)(623)}{1098} \qquad 75.27 = \frac{(174)(475)}{1098},$$

and

$$\chi^2 = \frac{(527-524.27)^2}{524.27} + \frac{(397-399.73)^2}{399.73} + \frac{(96-98.73)^2}{98.73} + \frac{(78-75.27)^2}{75.27}$$
$$= 0.207.$$

Using Table E with 1 df, we see that 0.207 is smaller than the smallest critical value given, so the P-value is greater than 0.25. There is no evidence of a difference in the proportion of adults who thought the military should have more control and the proportion of students at the University of Virginia who felt this way.

c)

Population	Sample size	Percent	Count
Student group 1	174	55%	96
Student group 2	199	16%	32

From the data, the estimates of the two proportions of students who succeeded are

$$\hat{p}_1 = 96/174 = 0.552$$
$$\hat{p}_2 = 32/199 = 0.161.$$

Next, we compute

$$\hat{p} = \frac{\text{count of successes in both samples combined}}{\text{count of observations in both samples combined}} = \frac{96+32}{174+199}$$

$$= 128/373 = 0.343.$$

The value of the z test statistic

$$z = \frac{\hat{p}_1 - \hat{p}_2}{\sqrt{\hat{p}(1-\hat{p})\left(\frac{1}{n_1}+\frac{1}{n_2}\right)}} = \frac{0.552-0.161}{\sqrt{0.343(1-0.343)\left(\frac{1}{174}+\frac{1}{199}\right)}} = 7.93.$$

The P-value is essentially 0, so there is strong evidence of a difference in the proportion who thought the military should have more control for the two different wordings of the question.

d)

	Yes	No	Total
Adults (1991)	702 675.37	222 248.63	924
Student Group 1	117 127.18	57 46.82	174
Student Group 2	129 145.45	70 53.55	199
Total	948	349	1297

The expected values are obtained as

$$675.37 = \frac{(924)(948)}{1297} \qquad 248.63 = \frac{(924)(349)}{1297}$$

$$127.18 = \frac{(174)(948)}{1297} \qquad 46.82 = \frac{(174)(349)}{1297},$$

$$145.45 = \frac{(199)(948)}{1297} \qquad 53.55 = \frac{(199)(349)}{1297}.$$

Next,

$$\chi^2 = \frac{(702-675.37)^2}{675.37} + \frac{(222-248.63)^2}{248.63} + \frac{(117-127.18)^2}{127.18} + \frac{(57-46.82)^2}{46.82}$$

$$= \frac{(129-145.45)^2}{145.45} + \frac{(70-53.55)^2}{53.55} = 13.85$$

Using Table E with $(3 - 1)(2 - 1) = 2$ df, we see that $0.0005 < \text{P-value} < 0.001$. There is strong evidence of a difference in the proportions who felt the news reports were being censored for the three groups. Both student groups were less likely than adults to believe that the military was censoring the news.

CHAPTER 14

INFERENCE FOR REGRESSION

SECTION 14.1

OVERVIEW

We first encountered regression in Chapter 3 . The assumptions that describe the regression model we use in this chapter are the following.

- We have n observations on an explanatory variable x and a response variable y. Our goal is to study or predict the behavior of y for given values of x.

- For any fixed value of x, the response y varies according to a normal distribution. Repeated responses y are independent of each other.

- The mean response μ_y has a straight-line relationship with x:

$$\mu_y = \alpha + \beta x$$

  The slope β and intercept α are unknown parameters.

- The standard deviation of y (call it σ) is the same for all values of x. The value of σ is unknown.

The **true regression line** is $\mu_y = \alpha + \beta x$ and says that the mean response μ_y moves along a straight line as the explanatory variable x changes. The parameters β and α are estimated by the slope b and intercept a of the least-squares regression line, and the formulas for these estimates are

$$b = r\frac{s_y}{s_x}$$

and

$$a = \bar{y} - b\bar{x}$$

in which r is the correlation between y and x, $\bar{y}$ is the mean of the y observations, s_y is the standard deviation of the y observations, $\bar{x}$ is the mean of the x observations, and s_x is the standard deviation of the x observations.

The **standard error about the least-squares line** is

$$s = \sqrt{\frac{1}{n-2}\sum \text{residual}^2} = \sqrt{\frac{1}{n-2}\sum (y-\hat{y})^2}$$

in which $\hat{y} = a + bx$ is the value we would predict for the response variable based on the least-squares regression line. We use s to estimate the unknown σ in the regression model.

A **level C confidence interval** for β is

$$b \pm t^* \text{SE}_b$$

in which t^* is the upper $(1 - C)/2$ critical value for the t distribution with $n - 2$ degrees of freedom and

$$\text{SE}_b = \frac{s}{\sqrt{\sum (x-\bar{x})^2}}$$

is the standard error of the least-squares slope b. SE_b is usually computed using a calculator or statistical software.

The **test of the hypothesis** $H_0: \beta = 0$ is based on the t statistic

$$t = \frac{b}{\text{SE}_b}$$

with P-values computed from the t distribution with $n - 2$ degrees of freedom. This test is also a test of the hypothesis that the correlation is 0 in the population.

GUIDED SOLUTIONS

Exercise 14.1

KEY CONCEPTS: Scatterplots, correlation, linear regression, residuals, standard error of the least-squares line

a) Sketch your scatterplot on the axes provided below.

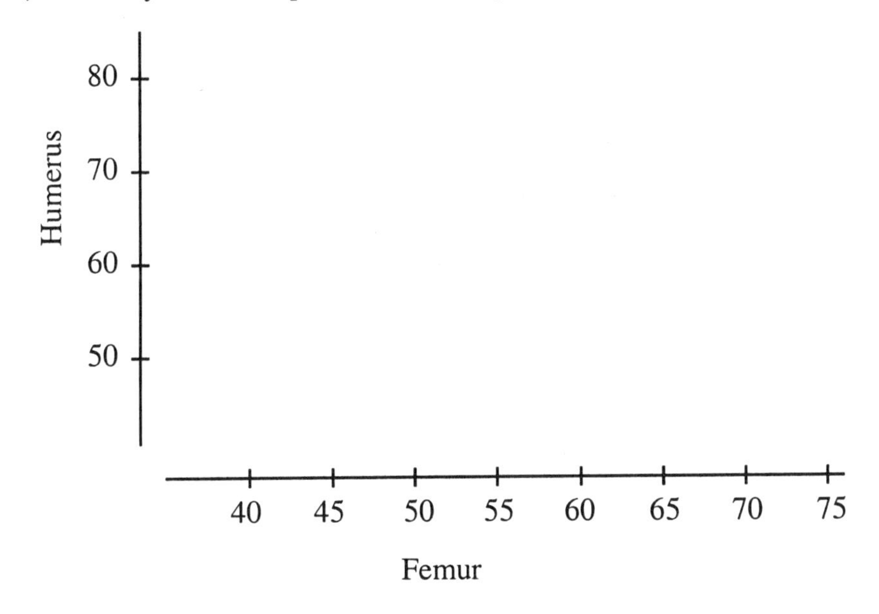

Next, use your calculator (or statistical software) to compute the correlation r and the equation of the least-squares regression line.

$r =$

Humerus length = __________ + __________ (femur length)

Do you think femur length will allow a good prediction of humerus length?

b) What does the slope β of the true regression line say about *Archaeopteryx*?

Enter your estimates of the slope β and intercept α of the true regression line in the space provided. Refer to your answer in part a for these estimates.

Estimate of $\beta =$

Estimate of $\alpha =$

c) To compute the residuals, complete the table.

Observed value of humerus length	Predicted value of humerus length : –3.65959 + 1.19690(femur length)	Residual (observed – predicted length)
41		
63		
70		
72		
84		
		Sum =

Next, estimate the standard deviation σ by computing

$\sum \text{residual}^2 =$

and then completing the following.

$$s = \sqrt{\frac{1}{n-2}\sum \text{residual}^2} =$$

Exercise 14.5

KEY CONCEPTS: Confidence interval for the slope

To determine the confidence interval, recall that a level C confidence interval for β is

$$b \pm t^* \mathrm{SE}_b$$

in which t^* is the upper $(1 - C)/2$ critical value for the t distribution with $n - 2$ degrees of freedom and

$$\mathrm{SE}_b = \frac{s}{\sqrt{\sum (x - \bar{x})^2}}$$

is the standard error of the least-squares slope b. In this exercise, b and SE_b can be read directly from the output given in Example 14.7 (look in the row labeled by the explanatory variable round 1). Their values are

$b =$

$\mathrm{SE}_b =$

Next, find t^* for a 95% confidence interval from Table C (what is n here?).

$t^* =$

Put all these pieces together to compute the 95% confidence interval.

$b \pm t^* \mathrm{SE}_b =$

Next, explain in plain language what the slope β means in this setting.

Exercise 14.6

KEY CONCEPTS: Least-squares regression line, tests for the slope.

a) Remember, the slope and intercept are given in the column labeled coef. Next, give the equation of the least-squares regression line by filling in the blanks below

Humerus length = ____________ + __________(femur length)

b) Recall that the test statistic is $t = \frac{b}{SE_b}$. Where do you find b and SE_b on the printout?

c) The number of degrees of freedom is $n - 2$. What is n? Next, use the appropriate row of Table C to estimate the P-value.

d) Refer to Example 14.6 in the textbook for the appropriate key strokes for the TI-83. What do you find with your calculator?

COMPLETE SOLUTIONS

Exercise 14.1

a) If we look at the data, we see that as the length of the femur increases, so does the length of the humerus. Thus, there is a positive association between femur and humerus length. A scatterplot of the data with femur length as the explanatory variable follows.

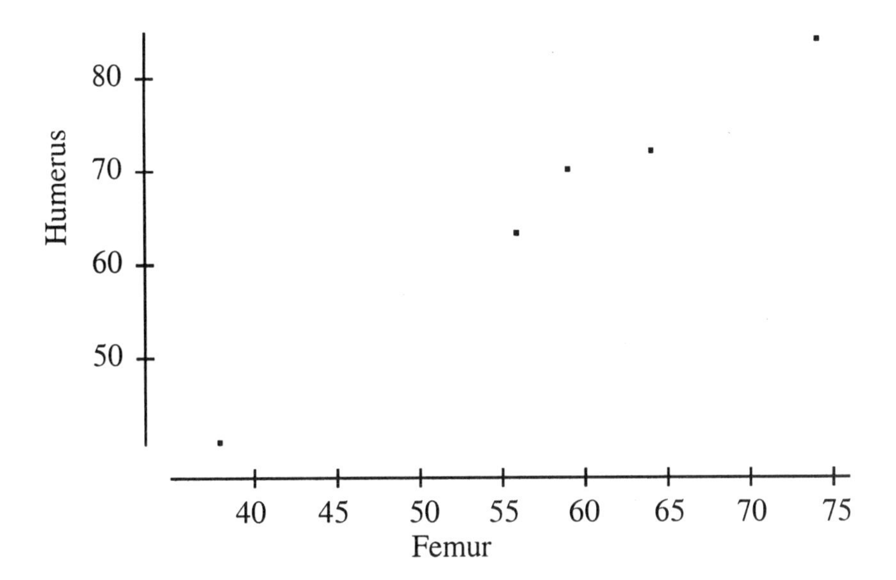

The scatterplot indicates a fairly strong positive association between femur and humerus length. If we calculate the correlation r and the equation of the least-squares line, we obtain the following.

$$r = 0.994$$

$$\text{Humerus length} = -3.65959 + 1.19690 \text{ (femur length)}$$

The correlation is rather high, so one would expect that femur length would allow good prediction of humerus length.

b) The slope β of the true regression line tells us the mean increase (in cm) in the length of the humerus associated with a 1-cm increase in the length of the femur in *Archaeopteryx*. From the data, the

$$\text{Estimate of } \beta = 1.19690$$

the slope of the least-squares regression line. From the data, the

$$\text{Estimate of } \alpha = -3.65959$$

the intercept of the least-squares regression line.

c) The residuals for the five data points are given in the following table.

Observed value of humerus length	Predicted value of humerus length: –3.65959 + 1.19690(femur length)	Residual (observed – predicted length)
41	41.822618	–0.822618
63	63.366820	–0.366820
70	66.957520	3.042480
72	72.942021	–0.942021
84	84.911022	–0.911022

The sum of the residuals listed is –0.000001, the difference from 0 due to rounding off. To estimate the standard deviation σ in the regression model, we first calculate the sum of the squares of the residuals listed:

$$\sum \text{residual}^2 = 11.785306$$

Therefore, our estimate of the standard deviation σ in the regression model is

$$s = \sqrt{\frac{1}{n-2}\sum \text{residual}^2} = \sqrt{\frac{1}{5-2}(11.785306)} = 1.982028$$

Exercise 14.5

We next note from Figure 14.5 that

$$b = 0.687747$$

$$\text{SE}_b = 0.23$$

For a 95% confidence interval from Table C with $n = 14$ (and $n - 2 = 12$),

$$t^* = 2.179$$

We put these pieces together to compute the 95% confidence interval.

$$b \pm t^*\text{SE}_b = 0.687747 \pm (2.179)(0.23) = 0.687747 \pm 0.50117.$$

The slope β of the true regression line tells us the mean increase in the score in round 2 associated with an increase of 1 stroke in round 1.

Exercise 14.6

a) From the printout, we see that the intercept has value –3.6596 and the slope value 1.1969. Thus, the equation of the least-squares regression line is

Humerus length = –3.6596 + 1.1969 (femur length).

b) The test statistic for testing H_0: $\beta = 0$ is the t statistic $t = \dfrac{b}{\mathrm{SE}_b}$. Using the results given on the printout, we see $b = 1.1969$ and $\mathrm{SE}_b = 0.0751$, so

$$t = \frac{b}{\mathrm{SE}_b} = \frac{1.1969}{0.0751} = 15.937.$$

c) t given in part b has a t distribution with $n - 2 = 5 - 2 = 3$ degrees of freedom. Using Table C (the row labeled 3), the upper tail probability for $t = 15.937$ is off the table and so must be less than 0.0005.

d) Using the key strokes described in Example 14.6 in the text, the TI-83 gives

intercept = 3.659586682

slope = 1.196900115

d.f. = 3

$t = 15.94050984$

P-value = 2.6842019×10^{-4}

which agrees with the results in parts a, b, and c.

SECTION 14.2

OVERVIEW

A **level C confidence interval for the mean response** μ_y when x takes the value x^* is

$$\hat{y} \pm t^* \mathrm{SE}_{\hat{\mu}}$$

in which $\hat{y} = a + bx$, t^* is the upper $(1 - C)/2$ critical value for the t distribution with $n - 2$ degrees of freedom and

$$\mathrm{SE}_{\hat{\mu}} = s\sqrt{\frac{1}{n} + \frac{(x^* - \bar{x})^2}{\sum (x - \bar{x})^2}}.$$

$\mathrm{SE}_{\hat{\mu}}$ is usually computed using a calculator or statistical software.

A **level C prediction interval for a single observation** on y when x takes the value x^* is

$$\hat{y} \pm t^* \mathrm{SE}_{\hat{y}}$$

in which t^* is the upper $(1 - C)/2$ critical value for the t distribution with $n - 2$ degrees of freedom and

$$\mathrm{SE}_{\hat{y}} = s\sqrt{1 + \frac{1}{n} + \frac{(x^* - \bar{x})^2}{\sum (x - \bar{x})^2}}.$$

$\mathrm{SE}_{\hat{y}}$ is usually computed using a calculator or statistical software.

GUIDED SOLUTIONS

Exercise 14.11

KEY CONCEPTS: Scatterplots, outliers and influential observations, r^2, confidence intervals for the slope, confidence intervals for the mean response

a) Use the axes provided to make your scatterplot.

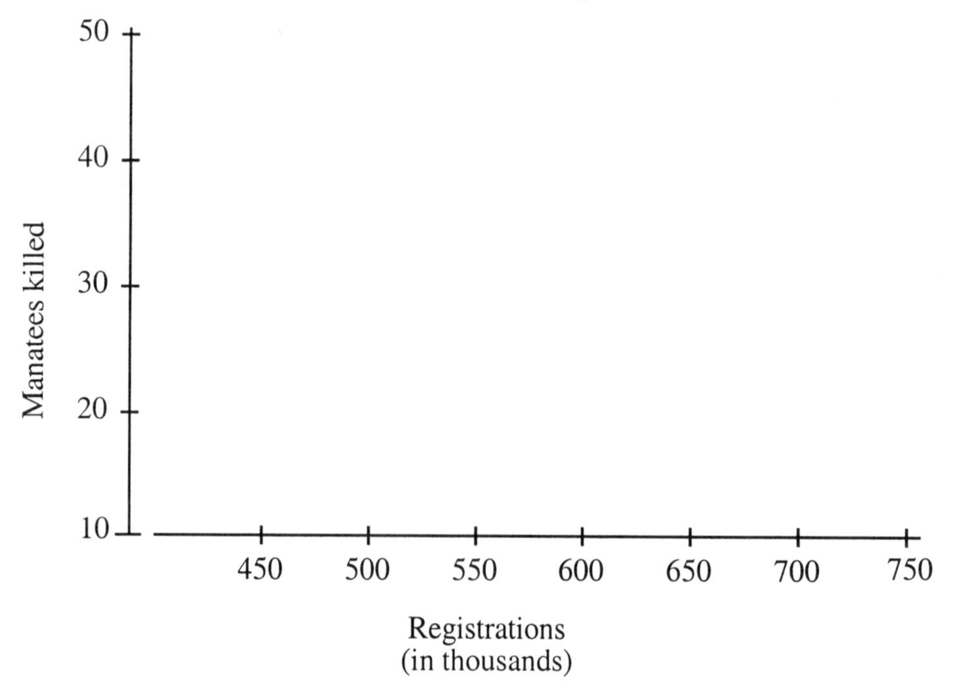

b) What pattern do you see in your plot? Are there any outliers or strongly influential points?

c) Refer to Section 3.3 if you need to review the meaning of r^2.

d) Explain what the slope β means in this setting.

To determine the confidence interval, recall that a level C confidence interval for β is

$$b \pm t^*\mathrm{SE}_b$$

in which t^* is the upper $(1 - C)/2$ critical value for the t distribution with $n - 2$ degrees of freedom and

$$\mathrm{SE}_b = \frac{s}{\sqrt{\sum (x - \bar{x})^2}}$$

is the standard error of the least-squares slope b. In this example, b and SE_b can be read directly from the output given in part c (look in the row labeled by the explanatory variable Boats). Their values are

$b =$

$\mathrm{SE}_b =$

Next, find t^* for a 90% confidence interval from Table C (what is n here?).

$t^* =$

Put all these pieces together to compute the 90% confidence interval.

$b \pm t^*\mathrm{SE}_b =$

e) Use the output given in part c to determine the equation of the least-squares regression line.

Manatees killed =

Next, use this equation to predict the number of manatees that will be killed in a year when 700,000 powerboats are registered. Remember, boats were measured in thousands of boats, so you should substitute 700, not 700,000, into the equation of the least-squares regression line.

Manatees killed =

f) Compare your answer in part a with the prediction of 45.97. If they differ, can you explain why?

Two 95% intervals are given in the output. Which of these is appropriate for predicting the mean number killed each year if Florida froze boat registrations at 700,000?

Exercise 14.12

KEY CONCEPTS: Confidence intervals for the mean response

A level C confidence interval for the mean response μ_y when x takes the value x^* is

$$\hat{y} \pm t^* \mathrm{SE}_{\hat{\mu}}$$

in which $\hat{y} = a + bx^*$, t^* is the upper $(1 - C)/2$ critical value for the t distribution with $n - 2$ degrees of freedom, and

$$\mathrm{SE}_{\hat{\mu}} = s\sqrt{\frac{1}{n} + \frac{(x^* - \bar{x})^2}{\sum (x - \bar{x})^2}}.$$

In this case, we are given $\mathrm{SE}_{\hat{\mu}}$ in part f of exercise 14.11 and do not need to compute it from the formula. What is the value of $\mathrm{SE}_{\hat{\mu}}$?

$\mathrm{SE}_{\hat{\mu}} =$

Next, complete the following. Note that $\hat{y}$ was computed in exercise 14.11.

$\hat{y}$ = predicted number of manatees killed for 700,000 powerboats

=

$n - 2 =$

t^* for a 90% confidence interval (refer to Table C) =

Therefore, the desired 90% confidence interval is

$\hat{y} \pm t^* \mathrm{SE}_{\hat{\mu}} =$

COMPLETE SOLUTIONS

Exercise 14.11

a) Here is a scatterplot showing the relationship between power boats registered and manatees killed. Because we are trying to use the number of powerboat registrations to explain the number of manatees killed, powerboat registrations is the explanatory variable.

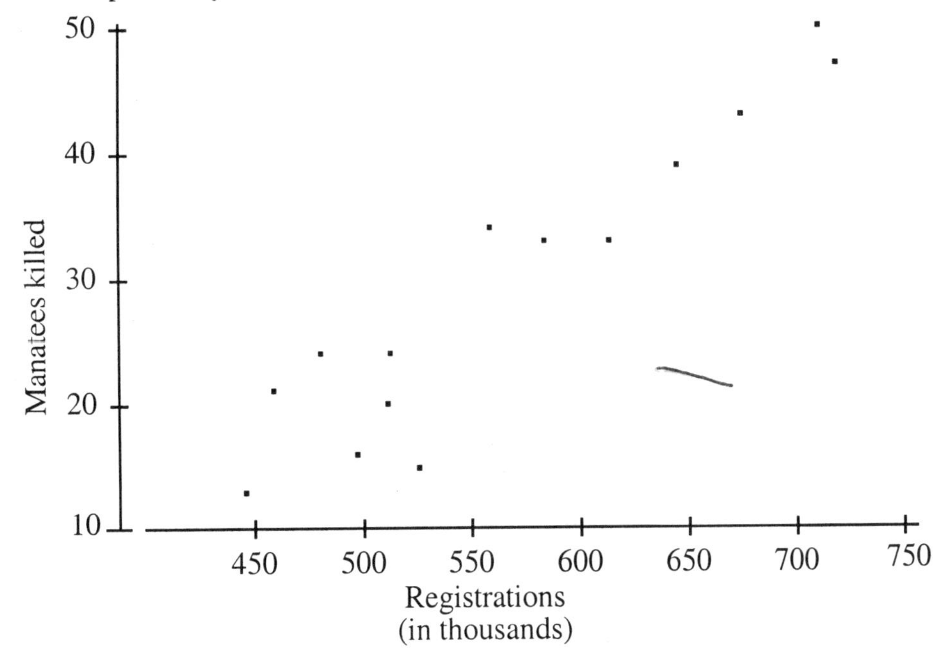

b) The overall pattern is roughly linear with a positive slope. There are no clear outliers or strongly influential data points.

c) We recall that r^2 tells us the percentage of the variation in the response variable accounted for by the explanatory variable in the least-squares regression line. In this case, $r^2 = 0.886$; thus, 88.6% of the variation in manatees killed is accounted for by powerboat registrations.

d) The slope β of the true regression line tells us the mean increase in the manatee deaths associated with an increase of 1000 powerboat registrations.

Next, we note from the output in part c that

$$b = 0.12486$$

$$SE_b = 0.01290$$

For a 90% confidence interval from Table C with $n = 14$ (and $n - 2 = 12$),

$$t^* = 1.782$$

We put these pieces together to compute the 90% confidence interval.

$$b \pm t^* SE_b = 0.12486 \pm (1.782)(0.01290) = 0.12486 \pm 0.02299$$

e) From the output given in part c, we see that the regression equation is

$$\text{Manatees killed} = -41.43 + 0.12486 \text{ Boats}$$

Plug in the value 700 for boats (boats were measured in thousands of boats when fitting the line, so we plug in 700, not 700,000) and the prediction is

$$\text{Manatees killed} = -41.43 + 0.12486(700) = 45.97.$$

f) The value 45.97 is the same as calculated in part e. We are interested in predicting the mean value (the mean number killed if Florida froze boat registrations at 700,000), so the appropriate interval is the 95% C.I. From the output the 95% confidence interval is thus

41.49 to 50.46.

Exercise 14.12

From the output in exercise 14.11 part f

$$SE_{\hat{\mu}} = 2.06.$$

In exercise 14.11, we also found

$$\hat{y} = \text{predicted number of manatees killed for 700,000 powerboats}$$

$$= -41.43 + 0.12486(700) = 45.97.$$

In addition, $n - 2 = 14 - 2 = 12$ so t^* for a 90% confidence interval = 1.782, therfore the desired 90% confidence interval is

$$\hat{y} \pm t^* \text{SE}_{\hat{\mu}} = 45.97 \pm 1.782(2.06) = 45.97 \pm 3.67.$$

SELECTED CHAPTER REVIEW EXERCISES

GUIDED SOLUTIONS

Exercise 14.15

KEY CONCEPTS: Scatterplots, r^2, tests for the slope of the least-squares regression line, prediction, prediction intervals

a) Use the axes to draw your scatterplot. We have placed % return-overseas on the vertical axis because this is the response variable.

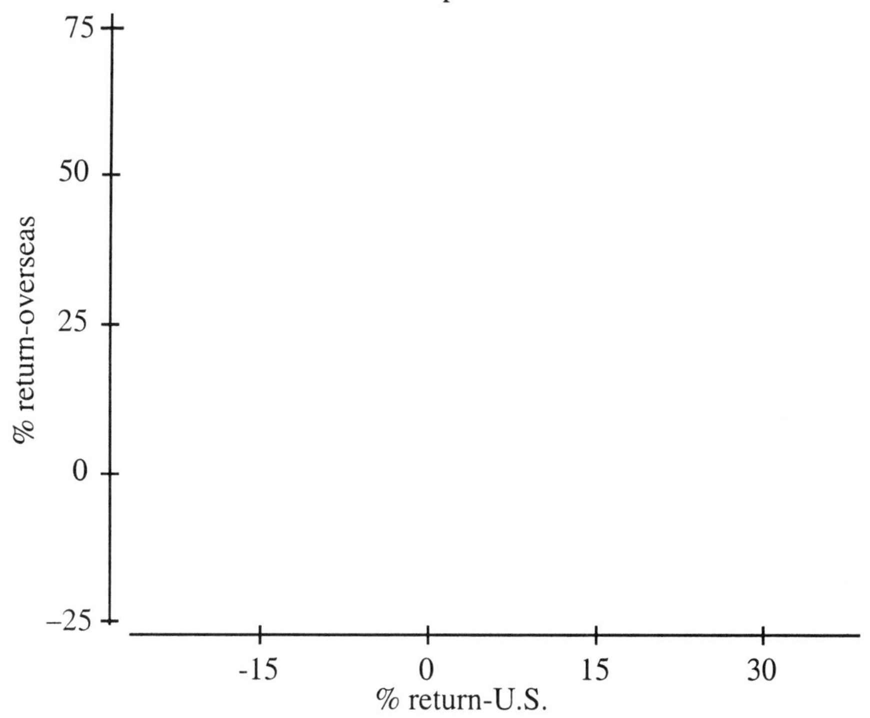

b) Use the r^2 value to help explain in words the relationship between U.S. and overseas returns.

Are there any clear outliers or strongly influential observations in your scatterplot?

c) First, state the hypotheses

H_0: H_a:

From the computer printout

$t =$

P-value =

How strong is the evidence for a linear relationship between U.S. and overseas returns?

d) Based on the output in part b, the equation of the least-squares regression line is

% return-overseas =

Next, use this equation to find the overseas % return when the U.S. % return is 20.

e) How does the "Fit" given in the output compare with your answer to part d?

Two 95% intervals are given in the output. Which is the appropriate interval for the return on foreign stocks next year if U.S. stocks return 20%?

Do you think the regression prediction is useful in practice?

Exercise 14.17

KEY CONCEPTS: Examining residuals

a) Use the axes to make your residual plot.

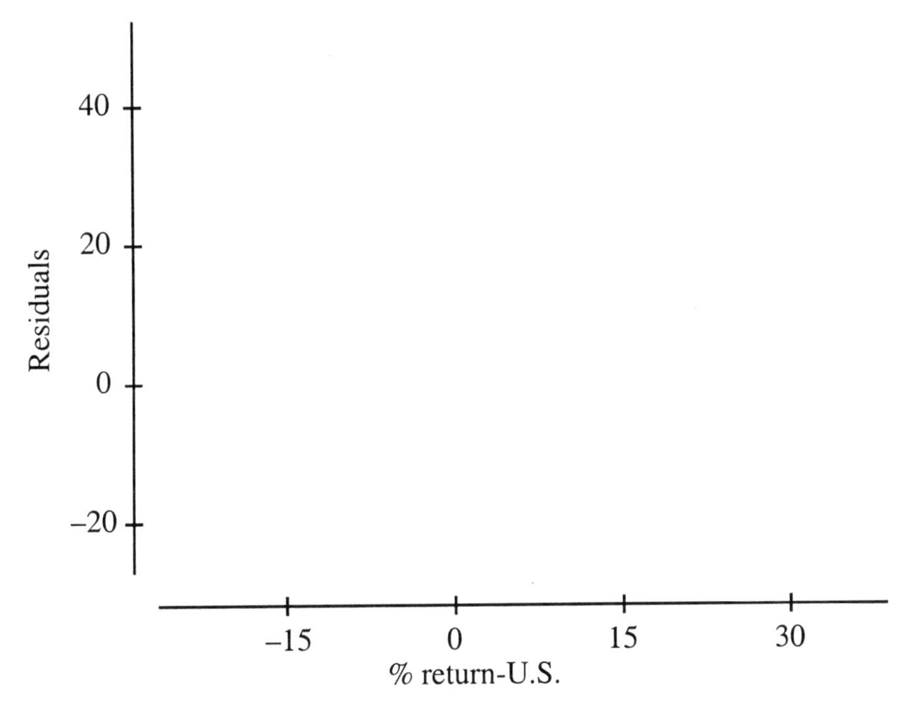

What do you see?

b) Make either a stemplot or histogram of the residuals.

Describe the shape of your plot. In what way is it somewhat nonnormal?

Identify the outlier. To what year does it correspond (look at the original data in exercise 14.15)?

COMPLETE SOLUTIONS

Exercise 14.15

a) Here is the desired scatterplot.

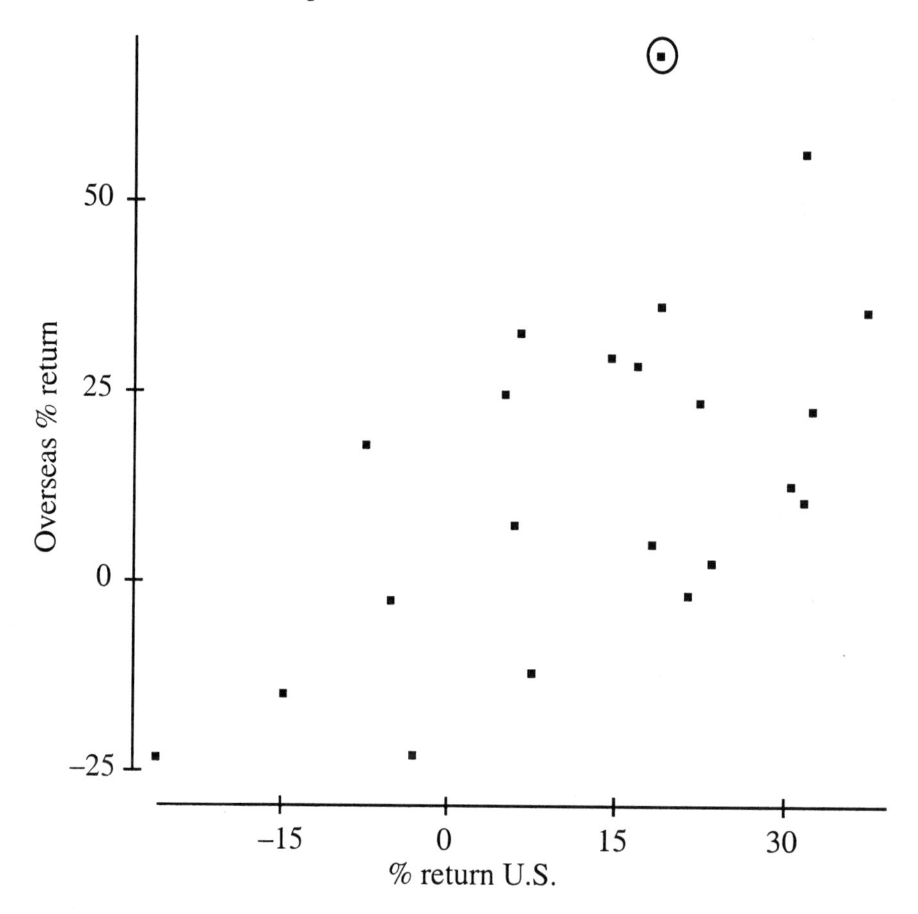

b) The value of r^2 is only 0.324. While the scatterplot shows a positive linear trend, only 32.4% of the variation in the return on foreign investments is explained by the least-squares regression line using the return on U.S. stocks as a predictor. Because 67.6% of the variation is unexplained, we would not expect predictions using the least-squares regression line to be very accurate, hence useful.

The point circled in the scatterplot in part a is an outlier. There are no clearly influential observations.

c) We wish to test the hypotheses H_0: $\beta = 0$ and H_a: $\beta \neq 0$.

From the output given, the test statistic is

$$t = \frac{b}{\mathrm{SE}_b} = 3.09$$

and the P-value is 0.006. The evidence appears to be strong for a linear relationship between U.S. and overseas returns.

d) Based on the output in part b, the equation of the least-squares regression line is

% return-overseas = 4.777 + 0.8130 (% return-U.S.).

If we substitute into this equation % return-U.S. = 20, we get

% return-overseas = 4.777 + 0.8130 (20) = 21.037.

e) The output agrees with the value we calculated in part d.

Two 95% intervals are given in the output. Because we are predicting a single value (the value in a year when U.S. stocks return 20%), we use the 95% prediction interval (P.I.). According to the output, this interval is

–21.97 to 64.04.

This is rather wide and not useful in practice. This is not surprising in light of the value of r^2 as discussed in part a.

Exercise 14.17

a) Here is a plot of the residuals against U.S. % return.

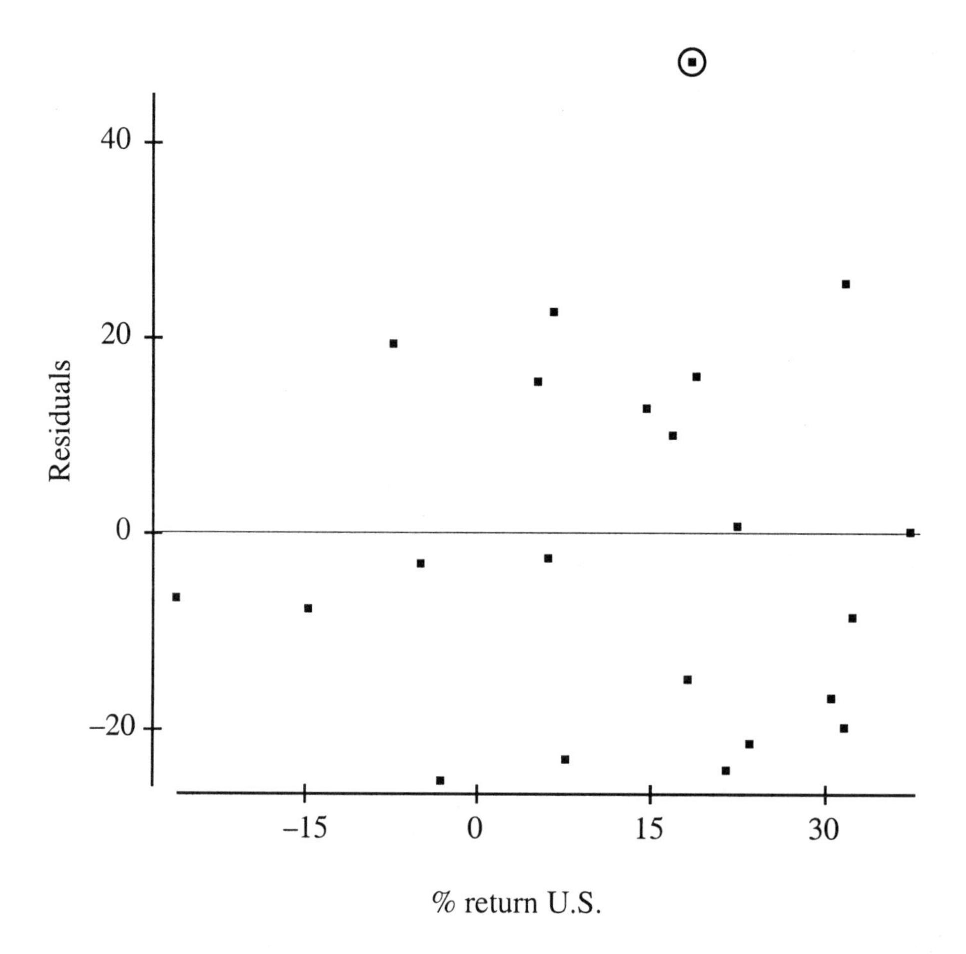

There is a modest outlier (the point circled in the plot), suggesting a moderate violation of the assumption of normality.

b) Here is a histogram of the residuals (you might also try making a stemplot).

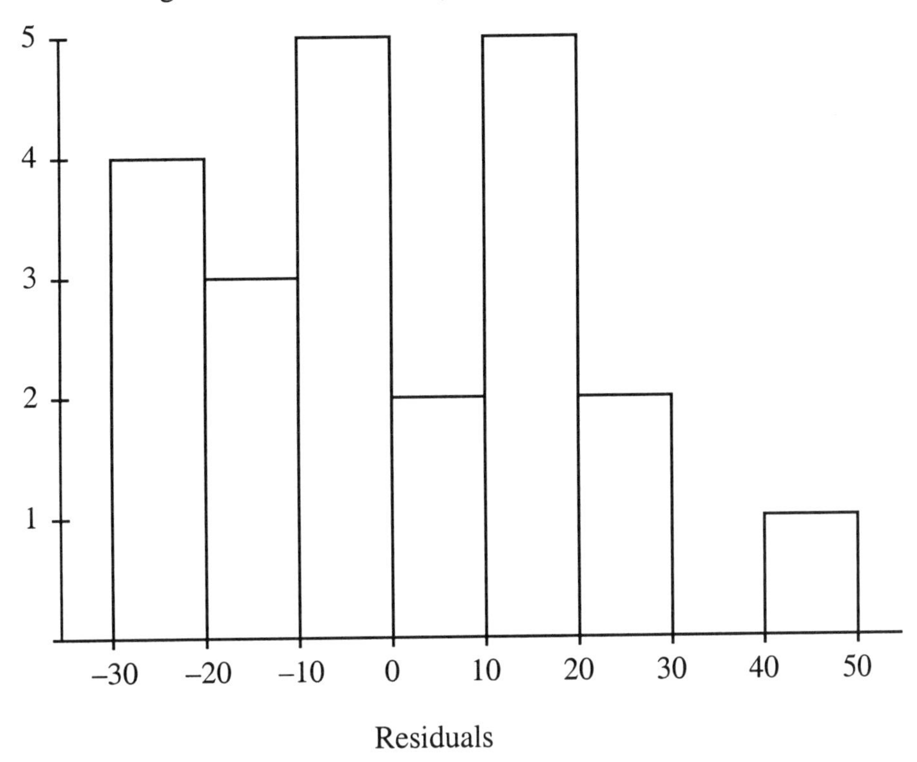

The shape is slightly skewed to the right. We see a possible outlier, the residual with value 49.5 (circled in the scatterplot in part a). If one examines the original data in exercise 14.15, this point corresponds to the year 1986 when the return on foreign stocks was unusually large.

CHAPTER 15

ANALYSIS OF VARIANCE

SECTION 15.1

OVERVIEW

There are formal inference procedures to compare the standard deviations of two normal populations as well as the two means. The validity of the procedures is seriously affected when the distributions are nonnormal, and they are not recommended for regular use. The procedures are based on the ***F* statistic**, which is the ratio of the two sample variances

$$F = \frac{s_1^2}{s_2^2}.$$

If the data consist of independent simple random samples of sizes n_1 and n_2 from two normal populations, then the F-statistic has the F-distribution, $F(n_1 - 1, n_2 - 1)$, if the two population standard deviations σ_1 and σ_2 are equal.

Critical values of the F-distribution are provided in Table D. Because of the skewness of the F-distribution, when carrying out the two-sided test, we take the ratio of the larger to the smaller standard deviation, which eliminates the need for lower critical values.

GUIDED SOLUTIONS

Exercise 15.7

KEY CONCEPTS: The F test for equality of the standard deviations of two normal populations

The data are from exercise 11.51, and are measurements on VOT (voice onset time) for a sample of children and adults. The summary statistics for the comparison of children and adult VOT scores are reproduced below.

Group	n	$\bar{x}$	s
Children	10	–3.67	33.89
Adults	20	–23.17	50.74

We are interested in testing the hypotheseses $H_0: \sigma_1 = \sigma_2$ and $H_a: \sigma_1 \neq \sigma_2$. The two-sided test statistic is the larger variance divided by the smaller variance, and if σ_1 and σ_2 are equal, has the $F(n_1 - 1, n_2 - 1)$. Remember that n_1 is the numerator sample size. To organize your calculations, first compute the three quantities below.

$$F = \frac{\text{larger } s^2}{\text{smaller } s^2} =$$

Numerator df $(n_1 - 1)$ =

Denominator df $(n_2 - 1)$ =

Compare the value of the F you computed to the critical values given in Table D, making sure to go to the row and column corresponding to the appropriate degrees of freedom, or as close as you can get to these degrees of freedom. Between which two critical values does F lie? What can be said about the P-value from the table?

COMPLETE SOLUTIONS

Exercise 15.7

a) Because the adults have the larger standard deviation (and variance), this is the variance that goes in the numerator, so that the degrees of freedom are $n_1 - 1 = 20 - 1 = 19$, and $n_2 - 1 = 10 - 1 = 9$.

To compute the value of the test statistic, note that it is the standard deviations and not the variances that are given as summary statistics. The standard deviations must first be squared to get the two variances. The larger standard deviation is for the adults and the sample variance for this group is $(50.74)^2 = 2574.5476$. For the the children, the sample variance is $(33.89)^2 = 1148.5321$. The value of the test statistic is

$$F = \frac{\text{larger } s^2}{\text{smaller } s^2} = \frac{2574.5476}{1148.5321} = 2.24.$$

Because the degrees of freedom (19, 9) are not in the table, we go to the table with the closest values (15, 9) or (20, 9) degrees of freedom. For these, we find that the smallest tabled critical values are 2.34 and 2.30, respectively, corresponding to a significance level of 0.10. The computed value of F is smaller than both of these. Because the test is two-sided, we double the significance level and conclude that the P-value is greater than 0.20. The data do not show significantly different spreads. Using computer software, we find that the P-value is 0.2156.

SECTION 15.2

OVERVIEW

The two-sample t procedures compare the means of two populations. However, when the mean is the best description of the center of a distribution, we may want to compare several population means or several treatment means in a designed experiment. For example, we might be interested in comparing the mean weight loss by dieters on three different diet programs or the mean yield of four varieties of green beans.

The method we use to compare more than two population means is the **analysis of variance (ANOVA) *F* test**. This test is also called the **one-way ANOVA**. The ANOVA F test is an overall test that looks for any difference between a group of I means. The null hypothesis is H_0: $\mu_1 = \mu_2 = \ldots = \mu_I$, in which we tell the population means apart by using the subscripts 1 though I. The alternative hypothesis is H_a: not all the means are equal. In a more advanced course, you would study formal inference procedures for a follow-up analysis to decide which means differ and to estimate how large the differences are. Note that formally the ANOVA F test is a different test from the F test described in Section 10.1 that compared the standard deviations of two populations. However, the ANOVA F test does involve the comparison of two measures of variation.

The ANOVA F test compares the variation among the groups to the variation within the groups through the ***F* statistic**,

$$F = \frac{\text{variation among the sample means}}{\text{variation among the individuals in the same sample}}.$$

The important thing to take away from this chapter is the rationale behind the ANOVA F test. The particulars of the calculation are not as important because software usually calculates the numbers for us.

The F statistic has the F distribution. The distribution is completely defined by its two degrees of freedom parameters, the numerator degrees of freedom and the denominator degrees of freedom. The numerator has $I - 1$ degrees of freedom, where I is the number of populations we are comparing. The denominator has $N - I$ degrees of freedom, where N is the total number of observations. The F distribution is usually written $F(I - 1, N - I)$.

The assumptions for ANOVA are that

- There are I independent SRSs.
- Each population is normally distributed with its own mean, μ_i.
- All populations have the same standard deviation, σ.

The first assumption is the most important. The test is robust against nonnormality, but it is still important to check for outliers and/or skewness that would make the mean a poor measure of the center of the distribution. As for the assumption of equal standard deviations, make sure that the largest sample standard deviation is no more than twice the smallest standard deviation.

Although it is generally best to leave the ANOVA computations to statistical software, seeing the formulas sometimes helps one to obtain a better understanding of the procedure. In addition, there are times when the original data are not available, and you have only the group means and standard deviations or standard error. In these instances, the formulas described here are required to carry out the ANOVA F test.

The F statistic is $F = \dfrac{\text{MSG}}{\text{MSE}}$, in which MSG is the **mean square for groups**,

$$\text{MSG} = \frac{n_1(\bar{x}_1 - \bar{x})^2 + n_2(\bar{x}_2 - \bar{x})^2 + \ . \ . \ . \ + n_I(\bar{x}_I - \bar{x})^2}{I - 1}$$

with

$$\bar{x} = \frac{n_1\bar{x}_1 + n_2\bar{x}_2 + \ldots + n_I\bar{x}_I}{N}$$

and MSE is the **error mean square**,

$$\text{MSE} = \frac{s_1^2(n_1 - 1) + s_2^2(n_2 - 1) + . \ . \ . + s_I^2(n_I - 1)}{N - I}.$$

Because MSE is an average of the individual sample variances, it also is called the **pooled sample variance**, written s_p^2, and its square root, $s_p = \sqrt{\text{MSE}}$, is called the **pooled standard deviation**.

We also can make a confidence interval for any of the means by using the formula $\bar{x}_i \pm t^* \frac{s_p}{\sqrt{n_i}}$. The critical value is t^* from the t distribution with $N - I$ degrees of freedom.

GUIDED SOLUTIONS

Exercise 15.9

KEY CONCEPTS: Stemplots, comparing means, interpreting ANOVA output

a) Complete the stemplots provided on the next page (they use split stems). From the stemplots, would you say that any of the groups show outliers or extreme skewness? You may want to review the material on stemplots in Chapter 1 if you have forgotten the details.

b) Give the five number-summary for each species.

	Minimum	Q_1	Median	Q_3	Maximum
Bream					
Perch					
Roach					

What do the data show about the weights of the three species?

Bream	Perch	Roach
12	12	12
12	12	12
13	13	13
13	13	13
14	14	14
14	14	14
15	15	15
15	15	15
16	16	16
16	16	16
17	17	17
17	17	17
18	18	18
18	18	18
19	19	19
19	19	19
20	20	20
20	20	20

Exercise 15.10

KEY CONCEPTS: ANOVA hypotheses, interpreting ANOVA output

a) State the null hypothesis being tested by the ANOVA F statistic test in both words and symbols.

H_0:

b) From the output, determine the values of the ANOVA F statistic and its P-value.

F statistic = P-value =

c) Using the results from exercise 15.9 as well as the ANOVA table, what do you conclude? You might want to look at the confidence intervals for the individual means as well.

Exercise 15.12

KEY CONCEPTS: ANOVA degrees of freedom, computing P-values from Table D

a) In the table, fill in the numerical values and explain in words the meaning of each symbol we are using in the notation for the one-way ANOVA. For consistency, let group 1 be Bream, group 2 be Perch and group 3 be Roach. In the calculations, remember that the largest perch was removed before doing the ANOVA.

Symbol	Value	Verbal meaning
I		
n_1		
n_2		
n_3		
N		

b) Use the text formulas and the results from part a to give the numerator and denominator degrees of freedom.

Numerator degrees of freedom =

Denominator degrees of freedom =

c) The value $F = 29.92$ needs to be referred to an $F(2, 107)$ distribution. Use denominator degrees of freedom equal to 100 to be conservative, because the entry for 107 is not in the table. What can you say about the P-value from Table D?

Exercise 15.18

KEY CONCEPTS: Comparison between means, checking assumptions, interpreting results from an ANOVA

a) By comparing the group means, what is the relationship between marital status and salary?

b) The ratio of the largest to the smallest standard deviations is

$$\frac{\text{largest sample standard deviation}}{\text{smallest sample standard deviation}} =$$

Does this allow the use of the ANOVA *F* test?

c) Calculate the degrees of freedom for the ANOVA *F* test by first computing *N* and *I*.

$N =$ $I =$

Numerator df = Denominator df =

d) The large sample sizes (particularly for the married men) indicate that the margins of error for the sample means will be very small, much smaller than the observed differences in the means. What is the numerical value of the standard error for married men? How does it compare with the differences in mean salaries between married men and the other groups? What does this imply about the ANOVA *F* test?

e) Is this an observational study or an experiment? How does that affect the type of conclusions that can be made?

Exercise 15.20

KEY CONCEPTS: ANOVA computations, pooled standard deviatión

a) It is easiest to use the sample sizes and the standard deviations from Figure 15.4, or you can recompute these from the data in Table 15.2. MSE is obtained from the formula (remember to square the standard deviations)

$$\text{MSE} = \frac{s_1^2(n_1 - 1) + s_2^2(n_2 - 1) + \ldots + s_I^2(n_I - 1)}{N - I} =$$

and the pooled standard deviation s_p is the square root of this quantity.

b) The formula for the confidence interval for the mean weight of perch is

$$\bar{x}_{\text{perch}} \pm t^* \frac{s_p}{\sqrt{n_{\text{perch}}}} =$$

in which the mean and sample size for perch are given in Figure 15.4, the pooled standard deviation was computed in part a and t^* is the appropriate critical value from Table C. What are the degrees of freedom for t^*?

Exercise 15.21

KEY CONCEPTS: ANOVA computations, mean square for groups

You will need the means and sample sizes from Figure 15.4. To compute MSG, you first need to compute the overall mean

$$\bar{x} = \frac{n_1\bar{x}_1 + n_2\bar{x}_2 + \ldots + n_I\bar{x}_I}{N} =$$

and then substitute the means, sample sizes, and overall mean into the formula

$$\text{MSG} = \frac{n_1(\bar{x}_1 - \bar{x})^2 + n_2(\bar{x}_2 - \bar{x})^2 + \ . \ . \ . \ + n_I(\bar{x}_I - \bar{x})^2}{I - 1} =$$

To complete the calculations, use the MSE from exercise 15.20 to give

$$F = \frac{\text{MSG}}{\text{MSE}} =$$

and find the P value as the area to the right of the computed value of F. What are the degrees of freedom?

COMPLETE SOLUTIONS

Exercise 15.9

a) Side-by-side stemplots are given below. All distributions are reasonably symmetric. The distribution for perch shows one extremely high outlier corresponding to the weight of 20.9 grams.

Bream		Perch		Roach	
12	0	12		12	
12		12		12	
13	333344	13	2	13	3
13	5677788889	13	69	13	6799
14	111233	14	3	14	0013
14	78899	14	55667889	14	677
15	001113	15	000000011112344	15	122344
15	5	15	56788999	15	6
16		16	0112333	16	1
16		16	8	16	
17		17	003	17	
17		17	56667789	17	
18		18	1	18	
18		18		18	
19		19		19	
19		19		19	
20		20		20	
20		20	9	20	

b) The five-number summary for each species are given below. Using the five-number summaries and the stemplots, we see that the distribution of weights of Bream and Roach are quite similar, with the Bream being slightly smaller. The spread in both distributions are about the same. The Perch are larger than either of the other two species, and the distribution is more spread out, even disregarding the outlier.

	Minimum	Q_1	Median	Q_3	Maximum
Bream	12.0	13.60	14.10	14.90	15.5
Perch	13.2	15.00	15.55	16.55	20.9
Roach	13.3	13.95	14.65	15.25	16.1

Exercise 15.10

a) In words, the null hypothesis is that the average weights of the three species in the lake are equal. In symbols, this corresponds to

$$H_0: \mu_{\text{Bream}} = \mu_{\text{Perch}} = \mu_{\text{Roach}},$$

in which μ_{Bream}, μ_{Perch}, and μ_{Roach}, are the true average weights of each of the three species in the lake.

b) Reading from the output, we have $F = 29.92$ and $P < 0.001$ (because the P-value is given only to three decimals, this is all we can say).

c) The low P value indicates differences in the mean weights of the three species, but does not tell us which species are different or in what direction. However, it is clear from the stemplots and the fact that the confidence interval for the mean weight of perch is completely to the right of the other two, that the mean weight of perch in the lake is higher than for the other two species. The stemplots show little difference between the weights of bream and roach, and the confidence intervals for the mean weight of these two species have considerable overlap.

Exercise 15.12

a) The table gives the value and states in words the meaning of each symbol used in a one-way ANOVA.

Symbol	Value	Verbal meaning
I	3	Number of groups
n_1	35	Number of Bream caught
n_2	55	Number of Perch caught (excluding the outlier)
n_3	20	Number of Roach caught
N	110	Total number of fish in the experiment

b) The ANOVA F statistic has the F distribution with $I - 1 = 3 - 1 = 2$ degrees of freedom in the numerator and $N - I = 110 - 3 = 107$ degrees of freedom in the denominator.

c) The critical value of 7.41 corresponds to a tail probability of 0.001 for an $F(2, 100)$ distribution. Because the value $F = 29.92$ exceeds this, we can say that the P-value is less than 0.001, which agrees with the computer output.

Exercise 15.18

a) The most obvious feature is that men who are or have been married earn more, on average, than single men. Men who are or have been married earn about the same amount, although divorced men appear to earn a little less, on average, than married or widowed men.

b) The ratio of the largest to the smallest standard deviations is

$$\frac{\text{largest sample standard deviation}}{\text{smallest sample standard deviation}} = \frac{8119}{5731} = 1.42 < 2$$

Because this ratio is less than 2, the sample standard deviations allow the use of the ANOVA F test.

c) We calculate

I = number of populations we wish to compare
= number of different marital statuses
= 4

n_1 = sample size from population 1 (single men) = 337
n_2 = sample size from population 2 (married men) = 7730
n_3 = sample size from population 3 (divorced men) = 126
n_4 = sample size from population 4 (widowed men) = 42

N = total sample size = sum of the n_i = 8235.

The degrees of freedom for the ANOVA F statistic are

$$I - 1 = 4 - 1 = 3$$

degrees of freedom in the numerator and

$$N - I = 8235 - 4 = 8231$$

degrees of freedom in the denominator.

d) The large sample sizes (particularly for the married men) indicate that the margins of error for the sample means will be rather small, much smaller than the observed differences in the means. For example, the standard error for the mean of married men is

$$\frac{\$7159}{\sqrt{7730}} = \$81.4$$

which is rather small compared with the difference of, for example, more than $5000 in mean salaries with single men (you can check that the standard error for single men is $312.2, also much smaller than the difference in means of more than $5000).

e) The differences in means probably do not mean that getting married raises men's mean incomes. This is an observational study, so it is not safe to conclude that the observed differences are due to cause and effect. A likely explanation of the observed differences is that the typical single man is much younger than the typical married man. As men get older, they are more likely to be married. Younger men have been in the firm for less time than older men and so will have lower salaries. This explanation may also explain the small differences between men who are or have been married. Married and widowed men (as compared to divorced men) are likely to include the most senior men in the firm. The most senior men are likely to have the highest salaries.

Exercise 15.20

a) $$\text{MSE} = \frac{s_1^2(n_1-1)+s_2^2(n_2-1)+...+s_I^2(n_I-1)}{N-I}$$

$$= \frac{0.770^2(34)+1.186^2(54)+0.780^2(19)}{107} = 1.006$$

which agrees with the computer output. The pooled standard deviation is $s_p = \sqrt{1.006} = 1.003$, which is given on the output.

b) From Figure 15.4, $\bar{x}_{\text{perch}} = 15.747$, $n_{\text{perch}} = 55$ and t^* is the 0.025 critical value of the t distribution with 107 df, or $t^* = 1.984$ (using df = 100 in Table C). Substituting these in the formula gives

$$\bar{x}_{\text{perch}} \pm t^* \frac{s_p}{\sqrt{n_{\text{perch}}}} = 15.747 \pm 1.984 \frac{1.003}{\sqrt{55}} = 15.747 \pm .268 = (15.479, 16.015).$$

Exercise 15.21

$$\bar{x} = \frac{n_1\bar{x}_1 + n_2\bar{x}_2 + ... + n_I\bar{x}_I}{N} = \frac{(35)(14.131)+(55)(15.747)+(20)(14.605)}{110}$$

$$= 15.0252.$$

because $N = 35 + 55 + 20 = 110$.

Substituting into the formula for MSG,

$$\text{MSG} = \frac{n_1(\bar{x}_1-\bar{x})^2 + n_2(\bar{x}_2-\bar{x})^2 + \ ...\ + n_I(\bar{x}_I-\bar{x})^2}{I-1} =$$

$$= \frac{35(14.131-15.0252)^2 + 55(15.747-15.0252)^2 + 20(14.605-15.0252)^2}{3-1}$$

$= 30.086.$

To complete the calculations,

$$F = \frac{\text{MSG}}{\text{MSE}} = \frac{30.086}{1.006} = 29.91,$$

which agrees well with the MINITAB output. The P-value is the area to the right of 29.91 for an $F(2, 107)$ distribution. Using the TI-83 or statistical software, we have P-value = 0.0000.